7 FUNDAMENTALS OF AN OPERATIONALLY EXCELLENT MANAGEMENT SYSTEM

Dr. Chitram Lutchman
Douglas Evans
Waddah Ghanem
Dr. Rohanie Maharaj

CRC Press
Taylor & Francis Group
Boca Raton London New York

CRC Press is an imprint of the
Taylor & Francis Group, an **informa** business

CRC Press
Taylor & Francis Group
6000 Broken Sound Parkway NW, Suite 300
Boca Raton, FL 33487-2742

Printed on acid-free paper
Version Date: 20141121

International Standard Book Number-13: 978-1-4822-0576-3 (Hardback)

Visit the Taylor & Francis Web site at
http://www.taylorandfrancis.com

and the CRC Press Web site at
http://www.crcpress.com

This book is dedicated to all the workers who work tirelessly on a daily basis to improve business performance. The book is especially dedicated to the health and safety professionals and technical subject matter experts who dedicate much of their daily lives to keeping us safe, both at work and during our private lives, in their continual quest of finding better ways to do things. To the many workers who have perished in tragic workplace incidents, and to the families and friends you may have left behind, we dedicate this knowledge to you.

Contents

Foreword

Following a series of catastrophic accidents in the petrochemical industry in the 1970s and 1980s (*Exxon Valdez*, Flixborough, Seveso, Bhopal, and Piper Alpha, to name a few), it became clear to regulatory bodies in the United States, Europe, and elsewhere that a fresh approach to safety was vital if recurrences of these incidents within the industry were to be averted. Investigation of these incidents indicated that the safety concepts that then existed in the form of prescriptive legislation and the reactive approach to the management of health and safety in industry were largely ineffective. Regulators failed to include measures that ensured the active participation of all employees in the creation of a safe working environment in the form of a positive safety culture.

In order to create this positive safety culture, major industries around the world carefully reviewed many safety management systems, and one of the systems that surfaced was process safety management (PSM). Success in creating a positive safety culture throughout an organization hinged on involving all employees and contractors in the management system. Industry realized that it was only by drawing on the vast and diverse knowledge and experience of all workers and by channeling this into a companywide atmosphere of safety awareness would success be achieved.

Through this approach, organizations were able to identify the strengths and weaknesses of their management systems. Once they were able to do that, they saw the need for a comprehensive safety management system aimed at the proactive identification, assessment, and control of hazards. The new approach or system necessarily embodied both process safety and personnel safety and was supported by means for measuring the quality and effectiveness of the various systems implemented by organizations. Efficient use of this new management system and its assurance of effectiveness are not possible if leadership accountability is not embedded and inherent in all the process elements. This resulted in the introduction of the PSM system in the early 1990s. Meanwhile, the concepts of an operational excellence management system (OEMS) were being pursued by ExxonMobil in response to the *Exxon Valdez* major incident. OEMS concepts became more apparent and pronounced in the early 2000s and were pursued in earnest by some of the leading global oil and gas producers. Insufficient and inadequate practical reference materials are among the primary conditions that prevented organizations from appreciating the competitive advantage they could potentially gain from integrating their environmental health and safety (EHS) and reliability systems under the OEMS umbrella.

This book provides a practical reference source of OEMS implementation materials derived from the authors' many years of experience, and

I am grateful for the opportunity to provide the Foreword for this book. As a corporate EHS leader with more than 30 years of experience in the field, I endorse this book as a wealth of practical and industry applicable processes and concepts for executing and sustaining an OEMS. The principles discussed in this book are sound, easy to follow, industry relevant, scalable, and practical. This book is an invaluable resource for organizations seeking continuous improvements and world-class EHS and business performance.

Having participated in the implementation of OEMS across the Bahrain Petroleum Company (BAPCO), I find that this book proves to be an exceptional compilation of processes and tools for implementing OEMS across any organization. The authors provide compelling evidence that OEMS is an imperative venture, particularly in a highly competitive, high-risk business environment characterized by dwindling nonrenewable energy sources and the relentless challenge to improve efficiency, reliability, EHS performance, and value maximization for stakeholders.

The authors highlight the importance of corporate governance and leadership in setting the tone required for the cultural shifts essential in creating an OEMS. Emphasis is placed upon the board of directors and executive leadership levels to internalize OEMS and translate that into real support to EHS improvements when creating OEMS. In this book, EHS is recognized as a core component of OEMS, and governance and the leadership are its foundations. OEMS principles lead to high-reliability organizations (HROs) that are well versed in the knowledge provided in this book. Indeed, the OEMS principles presented in this book were applied at BAPCO, which led to sustained improvements in EHS performance over the past 3 years (even across a major turnaround), and today we are a much more reliable and efficient organization. We are now more disciplined, applying consistent and standardized business practices designed to enable us to do things the right way all the time.

Future generations will not enjoy the luxury of job stability that is currently inherent in the oil and gas industry, in which one can develop a 30-year career within the same organization. Such stability permits employees the opportunity to experience growth and development and to observe shifts in management systems. This book provides an excellent summary of the management system's evolution from customer-focused, fact-based decision making to somewhat segmented and compartmentalized total quality management (TQM) in the early 1970s, to the current-day focus on OEMS that promotes full integration among people, processes, systems, and leadership for success.

The book also demonstrates the business performance impact of OEMS applications across leading oil and gas organizations, such as ExxonMobil, Suncor Energy, BP, and Chevron. The seven fundamentals of an OEMS identified by the authors provide a sequential approach for developing and executing an OEMS across any organization. This is a how-to process for developing and sustaining the OEMS. The fundamentals of OEMS provided and detailed in this writing include the following:

1. Emphasizing leadership, commitment, and motivation. This establishes critical emphasis on the creation of a shared vision, and the application of transformational leadership behaviors for inspiring the hearts and minds of people to achieve the goals of discipline and excellence.
2. Identifying and executing applicable and relevant elements to achieve operational excellence in your business. Select what is right for the organization and focus business resources and attention on these elements to drive discipline and excellence.
3. Establishing the baseline. Know where you are relative to where you want to be and strive for improved performance all the time.
4. Following the plan–do–check–act (PDCA) model. This is fundamental to any management system and requires operational discipline and organizational compliance with this very simple but critical business process, which ensures continual improvements. Following the PDCA model allows organizations to manage work in closing gaps identified in the management system when the baseline is established.
5. Auditing for compliance and conformance. Audit the business on an ongoing basis in a collaborative, noninvasive, nonpunitive way relative to the management system requirements. This drives workers toward greater discipline and excellence.
6. Closing the gaps post-audit. This is an essential process in the path to operational excellence.
7. Making continuous improvements. The book introduces networks and communities of practices for generating and driving continuous improvements and shared learning in the business.

This book provides a valuable resource of OEMS workshop materials accessible to users for creating a shared vision of OEMS and for communicating OEMS and training workers across the organization. The workshop materials are essentially a valuable resource for senior leaders, managers and supervisors, and EHS and training providers in any organization who seek to create an OEMS. (The OEMS workshop materials can be accessed at http://www.safetyerudite.com.) This book provides an excellent opportunity for organizations to significantly enhance the spread and execution of OEMS within and across organizations.

Ahmed Khalil Ebrahim
Corporate Leader,
Fire, Safety, and Environment, BAPCO
Kingdom of Bahrain

Acknowledgments

Operational discipline and operational excellence, relatively new terms in the current business environment, are deemed essential for high and sustained business performance. In this book, the authors have articulated the requirement for developing and sustaining an operational excellence management system (OEMS) in any organization. The authors include prior and current employees of both Beyond Petroleum (BP) and Suncor Energy, Inc. We have used teachings from both of these organizations to influence the content of this book, which is therefore recognized accordingly. This work would not have been completed without the help and contributions of many indirect and direct contributors—peers, consultants, and spouses; they are also recognized for their contributions. The quality of this work would not have been of this standard without the influence of our Generation Y contributors. In view of this, the authors acknowledge the contributions of Darryl Kissoon for his user testing and grammatical reviews. Special thanks also to Kevan, Megan, and Alexa Lutchman, who each in turn performed grammatical and readability reviews. Their contributions helped to simplify this work, which is now presented in a manner suitable to all audiences. Once again, special thanks are also extended to Nishal Sankat for his online research contributions.

About the Authors

Dr. Chitram (Chit) Lutchman (Principal Author)
DBA—Doctor of Business Administration
MBA
Certified Safety Specialist (CSP)
Canadian Registered Safety Professional (CRSP)
1st Class Power Engineer
BSc Agriculture Sciences

Chitram Lutchman, a 2011 recipient of the prestigious McMaster Arch Award for unique and exceptional contributions to Canada, is an experienced safety professional with extensive frontline and leadership experience in the energy industry. Having sustained some disability from an industrial work accident, Dr. Lutchman developed a passion for improving health and safety in the workplace. With international oil and gas experience, he has experienced the two extremes of organizational health and safety practices and business performance. As an employee of Canada's largest oil and gas producer, he pioneered work aimed at improving contractor health and safety management within his organization. In prior roles, Dr. Lutchman worked as a corporate leader in loss management and emergency response. He also functioned in project management leadership roles in the commissioning and start up of Canada's first commercial SAGD and cogeneration facility in Fort McMurray, Alberta. He published his first book, *Project Execution: A Practical Approach to Industrial and Commercial Project Management*, in 2010. In 2012, he was the principal author of another book: *Safety Management: A Comprehensive Approach to Developing a Sustainable System*. He is also the principal author of this book and has coordinated the contributions of his coauthors to provide readers with this practical package of knowledge, which can contribute significantly to improving the health and safety of all workers, including new graduates, safety professionals, business leaders, and contractors. Dr. Lutchman is also passionate about the organized sharing of knowledge in the areas of HSSE, PSM, and management systems to ensure the health and safety of our most precious assets—people (in particular, Generation Y), while generating support for superior business performance for all stakeholders.

Douglas J. Evans
P.Eng.
BSc, Engineering
BSc, Chemical Engineering
P.Eng. APEGA

Douglas J. Evans is a graduate of Dalhousie University and the Technical University of Nova Scotia, where he earned a BSc and chemical engineering degrees in 1974 and 1976, respectively. He established his working career in the oil industry with Texaco Canada in Montreal, Quebec. During his employment with Texaco, Imperial Oil, Petro-Canada, and now Suncor, he has had an opportunity to experience Canada's cultural and geographic variety, with multiple work assignments in eastern and western Canada. Evans has held a variety of senior leadership roles in the areas of engineering, technology, strategic planning, supply, and operations management over his 38-year career. More recently, he was appointed general manager of technical services in Suncor's corporate offices after the merger with Petro-Canada. This role involves providing a center of excellence in process engineering and operations controls to support safe, reliable, and low-cost operations at all of Suncor's operating facilities and for enhanced business performance. His current focus is leveraging Suncor's Technical Excellence Networks to capture and share knowledge and best practices and to generate continuous improvements in the business. He is also responsible for stewarding process safety standards compliance and risk reduction across Suncor. Evans is currently a member of the AFPM Process Safety Workgroup, which is focused on enabling improved process safety performance in the North American refining industry. His passion for process safety management and risk reduction was also well utilized to drive deep organizational learning across Petro-Canada as a result of the BP Texas City tragedy. His previous roles have included director of process technology and reliability for Petro-Canada's Downstream refinery operations, director of business integration for Petro-Canada Downstream, operations director at the Petro-Canada Oakville refinery, area team manager of crude units and powerformer at Imperial's Strathcona refinery, and technical services manager at Strathcona Refinery. Evans has managed large expense and capital budgets and has been responsible for the safe and reliable operation of major refinery assets. What he enjoys most is working with talented and motivated people to achieve personal and business success. His reward is in supporting others to succeed.

Waddah Ghanem Al Hashmi
MBA
MS. Environmental Sciences
B.Eng. (Hons.) Environmental Engineering
Diploma Environmental Management
Diploma Safety Management
Fellow of the Energy Institute, United Kingdom
Associate Fellow of the Intuition of Chemical Engineers

Waddah Ghanem Al Hashmi is the chief EHSQ compliance officer for ENOC Group. He is chairman of various committees within the organization, including the Wellness & Social Activities Program, which serves over 6,000 employees. Ghanem is an environmental engineer who earned his degree from the University of Wales, College Cardiff. He began his career working as a consultant for Hyder Consulting Middle East and later transitioned into the first oil refinery in Dubai, UAE, ENOC Processing Company LLC (EPCL), during the construction and precommissioning phases. Ghanem holds a distinction-level MSc in environmental science from the University of the UAE. He earned an executive MBA through Bradford University in the UK, in which he specialized in organizational safety behavior. With extensive hands-on and frontline experience, he has broad knowledge of environmental management systems and pollution control, fire and safety compliance and design reviews, development and administration of occupational health management systems, EHS management systems auditing, job safety task analysis, total quality management, and other HSE capabilities. His contribution to the EHS field includes more than 70 presentations and technical papers at various local, regional, and international conferences and forums. Ghanem has been a member of the Executive Committee of the Emirates Safety Group and served as an adviser for the HCT Environmental Sciences Program Committee. He was a member of the Petroleum and Lube Specification Committee for the government of the UAE, and a member of the Demand Side Management Committee and HSE Committee at the Dubai Supreme Council of Energy. He is quite passionate about EHSQ issues, and his current doctoral research work with the Bradford School of Management, Bradford University, UK, is focused on operational corporate governance, EHS leadership, and compliance systems. He was also appointed as the vice chairman of the board of the Dubai Centre for Carbon Excellence (DCCE) in 2010.

Dr. Rohanie Maharaj
PhD Food Science and Technology
MSc and M.Phil. Food Technology
BSc Natural Sciences

Rohanie Maharaj is a PhD graduate in food science and technology from Université Laval in Canada. She has been employed at the University of Trinidad and Tobago (UTT) for the past 5 years as an associate professor in biosciences, agriculture, and food technologies (BAFT), and is the program leader for the BSc diploma and certificate programs in food technology. Dr. Maharaj is also trained in Six Sigma methodology and is a senior member of the American Society for Quality (ASQ) and the Institute of Food Technologists (IFT). She is a certified evaluator for the Accreditation Council of Trinidad and Tobago (ACTT), chair of the Food and Beverage Industry Development Committee (FBIDC), and currently serves on a number of internal UTT committees. She has more than 30 years of experience working in academia and industry, including 14 years in industry, with 12 of these spent at Johnson & Johnson leading quality and reliability. She was a regional director and member of the board of Johnson & Johnson Caribbean for 7 years. She has extensive industrial experience, having worked in various capacities, such as quality assurance/regulatory affairs director, process excellence director, operations director, compliance director, and quality assurance manager at Johnson & Johnson Caribbean facilities spanning Trinidad, Jamaica, Dominican Republic, and Puerto Rico. She received the Award of Excellence in 2005 at Johnson & Johnson for achieving exceptional business results. Dr. Maharaj was also instrumental in obtaining the ISO 14001 environmental certification for the Trinidad manufacturing facility in 2004. More recently, in 2013, she received the National Institute of Higher Education, Research, Science, and Technology (NIHERST) Award for Excellence in Science and Technology.

Section I

Introduction to Management Systems

1

Corporate Governance and EHS Leadership

The economic theory of *value maximization* as a proposition to the shareholder is deeply entrenched in over two centuries of economic theory and research. Starting with Adam Smith, economists have argued that both the social wealth and welfare of people is influenced by the economic success of corporations, which seek to maximize the stream of profits that can then be divided among their investors.

Today, this theory has been developed to cover a more long-term market value of corporations, as suggested by Chew and Gillan (2005). A competing theory is the *stakeholder theory*, in which the difference is that it looks at all the stakeholders and argues that instead of maximizing the returns to shareholders, the *total value* should be maximized for all the stakeholders, including employees, customers, suppliers, the local community, and the public. This brings about a different approach, and quite interestingly, this theory has been adopted by many organizations, professionals, and even governments (Jensen, 1993, as discussed in Chew and Gillan, 2005; Handy, 2002; Goodpaster and Matthews, 1982).

With the development and growth of both size and complexity of corporations, corporate governance and control has become increasingly important. The systems of corporate governance and control have also come under greater scrutiny in recent years, and organizational investors have demanded effective controls be put in place to ensure that their investments are not at risk of loss (Dunlop, 1998). However, equally as important has been the rapid change in more recent times, which Michael Jensen (1993, as cited in Chew and Gillan, 2005) explains has changed the economic landscape as rapidly as within the 19th-century Industrial Revolution. This, he argued, was mainly attributed to the vast and rapid changes in technology and organizational aspects that completely changed many models in production and labor markets. Kendall and Kendall (1998) explained that this phenomenon arose primarily because the size of companies increased, and with much higher production scales, shareholders ceased to manage these organizations and hired professional managers instead. A second contributing factor was the technological advances and accompanying gain from the economics of scale. As time went on, the professional managers moved to eventually become board members. This was because the organizations became bigger and more complex in their nature, and therefore the board member was seen as an important *advisor* to the shareholder.

As companies evolved over the years from small and medium-sized enterprises to today's larger corporations working in diverse businesses and in various geographical locations, logistics and other complexities required that a more sophisticated and greater deal of control systems be employed. As organizations expand, they demand a greater deal of planning, accounting, operational management, and systems in all their various functionalities (Leavy and McKiernan, 2009).

Such growth led to the development of more regimented systems with process mapping, procedures, and checklists. As Collins (2001) explained, these systems are usually employed in response to a senior person within the organization seeking to improve stewardship. Furthermore, a company's board of directors may also demand that better corporate monitoring and control be exercised in a hierarchical and systematized organization. In doing so, many organizations face the invertible, slow death of innovation, as many of the operations become regimented, making it necessary to follow strict guidelines and procedures that the burdens of bureaucracy bring about.

These systems generally come about to compensate for incompetence, inconsistencies (which can no longer be tolerated in certain high-risk industries such as aviation and oil and gas), and a lack of discipline. To have a great organization, Collins (2001) explained that a balance between a high ethic of entrepreneurship and a high culture of discipline is required. It may be considered significant that

> the good-to-great companies built a consistent system with clear constraints, but also gave people freedom and responsibility within the framework of that system. They hired self-disciplined people who didn't need to be managed, and then managed the system, not the people. (Collins, 2001, p. 125)

In this context, governance and control systems by their very nature are constraining, and establish certain requirements and expectations that have to be fulfilled by people who are then measured again against these performance criteria. In recent years, a greater amount of clarity has been demanded from organizational leadership and boards of directors. Zukis et al. (2010) explained that in the United States, with the internal control framework provided by the Committee of Sponsoring Organizations of the Treadway Commission (COSO), organizations are now leveraging this framework and its recommendations beyond financial reporting.

According to Zukis et al. (2010), the approach to policies and procedures now requires companies to

1. Develop a functional model identifying business processes (i.e. a Process Classification Framework) so that a well-defined policy and procedure framework exists.
2. Support the framework with established processes to continually evaluate, update, and communicate policy changes throughout the organization.
3. Leverage this framework consistently across the organization in support of various business processes. (p. 2)

PricewaterhouseCoopers (PWC) (2005) explained that with corporate governance where public and political pressure is mounting, many organizations in today's business environment are subject to an expectation of greater transparency and corporate responsibility, especially from senior management and boards of directors. Such expectations and proactive board responses in developing ethics subcommittees (sometimes also called supervisory committees) to oversee the behavior of the board members to ensure highest standards of compliance is driving better governance across many organizations.

Corporate governance has been defined differently by different countries. Kendall and Kendall (1998) explained that this is due primarily to the differences in the terminology and the concerns that different nations have regarding the reality of managing the corporation effectively. Kendall and Kendall (1998) suggested corporate governance included "to be seen to be acting responsibly, and informing all interested parties, or stakeholders, of decisions which will affect them" (pp. 18–19). McGregor (2000) offered a different definition, suggesting: "Governance is the process whereby people in power make decisions that create, destroy or maintain social systems, structures and processes" (p. 11).

Bain and Band (1996) did not provide an exact definition, but rather explained that it is about having the right standards working within the organization. In a survey conducted by Bain and Band (1996), respondents (managers) suggested many different definitions, including

> having an appropriate pay policy for senior members of the industry; providing checks and balances to avoid excesses of top bosses; having a set of procedures to protect the organization from fraud or loss due to poor practice; providing checks on the management thus protecting shareholders; curbing the worst excesses of a greedy managing class; providing a control climate suitable to the organization. (p. 3)

They predicted that "the new governance process is based on continuing dialogue and debate among key, long-term institutional and other investors about specific, substantive aspects of corporate policy" (p. 3).

Leavy and McKiernan (2009) defined corporate governance as "the process of serious decision-making at the controlling heart of the organization. For most practical purposes, this means the board and the CEO and the ultimate arbiters" (p. 46). The most appropriate definition of corporate governance, in the author's view, is provided by the Cadbury Committee, which defined corporate governance in a simplistic yet all-inclusive way: "Corporate governance is the system by which companies are directed and controlled" (p. 14).

Boards of directors are responsible for governing the organization they oversee. The shareholders' role in governance is to appoint the directors and the auditors and to guarantee that an appropriate governance structure exists in their quest for profit maximization (not always value maximization) or return on investments. The responsibilities of the board include setting the company's strategic goals, instituting leadership to achieve these goals, overseeing the leadership of the business, and reporting to shareholders on their stewardship. The board's actions are subject to laws, regulations, and the shareholders in general meetings (Cadbury Committee Report, 1992).

Chew and Gillan (2005) suggested that corporate governance can be described through its most important facet—organizational design and architecture. In their view, three key elements exist:

- The assignment of decision-making authority, i.e., who gets to make what decisions.
- Performance evaluation, i.e., how is the performance of employees and their business units measured?
- Compensation structure, i.e., how are employees (including senior managers) rewarded or penalized for performance?

McGregor (2000) explains that more and more business managers and leaders have the power to affect the quality of life of many people. Their decisions often profoundly impact the internal and external stakeholders, such as employees and the public, as well as the environment at large. Today, stakeholders are showing a growing appreciation for the softer issues of governance, as they seek to embrace and respond to the needs for social responsibility, empathy, and caring governance that are fundamental to understanding how organizations function, whether they are profit or nonprofit enterprises or even government agencies.

On the initial review of the various codes that have been developed in various countries (UK, Cadbury Code, 1992; Turnbull Report, 1999; United States, Sarbanes–Oxley Act, 2002; South Africa, King III, 2009), it is clear that the concept of corporate governance has grown in importance, with greater transparency being demanded by all stakeholders, including shareholders. Also, as demonstrated in South Africa, which applied this as an effective code of practices (see King I, King II, and King III), many of the South African companies have been able to enjoy a greater influx of foreign direct investment (FDI) with greater confidence of the investor.

There has been much discussion and debate in the United States between various institutions regarding the Sarbanes–Oxley (SOX) Act, which was passed soon after the Enron collapse/scandal. The act operates within the principle of *comply or else face the consequences*. The effectiveness of the act has been argued against considerably, with claims that are ineffective in adding a holistic value to the corporations. Some estimate that the cost of compliance to SOX in the United States amounts to a greater value than the total write-off amounts on Enron, WorldCom, and Tyco combined (King, 2009).

In 1992, the UK government commissioned the Cadbury Committee Report on Financial Corporate Governance of Companies. The report focuses on the role of the chairman, CEO, and directors (especially the role of independent directors). Much of the spirit of the debate that the committee went through reflects some of the aspects that relate to the nonfinancial decisions made by organizations (Clutterbuck and Waine, 1993).

The field of environmental health and safety (EHS) has grown in importance over the past three to four decades, mainly driven by various factors that include the impact of changing legislation on organizations, and the onus being placed firmly on organizations to protect their employees, contractors, and the public from adverse EHS impacts from their operations (HSE, 2001). In the UK, the Health and Safety Commission (HSC) has adopted the recommendations of the Turnbull Report (IoCA, 1999) on the internal control systems, and corporate governance provides a renewed focus on health and safety performance and the control of health and safety risks.

The importance of health and safety is directly related to principles of the prevention of loss. This is applicable in almost any business, no matter what operations are involved and regardless of the level of risk, but perhaps more significantly in the oil and gas and other high-risk/high-reliability industries where such accidents can lead to immense destruction to people and property, e.g., BP Texas refinery, with 15 fatalities, more than 170 injuries, and a cost to BP in both financial and reputation loss (Baker et al., 2007). The 2010 disaster off the Gulf of Mexico with the BP offshore operations as one of the most serious in terms of impact on the economy, environment, and people. As a consequence of that incident, the CEO was removed from his post for leadership shortcomings related to the failure to demonstrate safety leadership at the top of the organization (Bani Hashem, 2011).

Demonstrated safety leadership failures reside in the simple understanding that workers in any capacity require protection when it comes to their health and safety in the workplace. There is thus a legal aspect to health and safety at work. In the UK, legal protection to all (including employer, employee, contractors, the public, etc.) is provided through the Health and Safety at Work Act 1974 (HSAWA 1974). In Canada, Bill C-45 provides similar protection. The Occupational Safety and Health Act of 1970 in the United States catered to the right to a safe workplace. However, regardless of legislation with respect to health and safety under international common-law principles, a safe workplace is now a basic requirement from every employer, and all employers are required to provide reasonable care and protection for all workers, including contractors.

The management of risk concerns itself with the prevention of loss. Regulators provide negative reinforcement by issuance of enforcement, improvement, or prohibition notices from the local authority having jurisdiction for health and safety. Good loss prevention strategies also help organizations and individuals avoid punitive action in which criminal courts can impose fines, and compensatory lawsuits in civil courts, leading to compensation claims and imprisonment for breaches of legal duties. This can affect companies or individuals, and thus their operations and reputation.

Accidents are costly to organizations. The indirect costs of an incident are estimated to be as much as 30 times the direct losses incurred from an incident (DNV, 1996). The Health and Safety Executive in the UK estimates that for every £1 of insured loss, there is an estimated uninsured loss of between 8 and 36 times that which is insured (HSE, 1996). Insurances are often available to protect employers, and these include employers' liability, public liability, workers' compensation, and fire and perils. But is this the right way to address risk management?

It should be noted that losses associated with lost production time, loss of highly trained personnel, impacts on employee morale and productivity, and time and resources spent investigating the incident cannot always be recovered (BSC, 2005). Therefore, corporate governance, which makes loss management a priority, generally results in more sustainable performance. In the past 5 to 10 years, no company has felt the same crippling impact on its reputation (and shareholder confidence) and share price (company value) as BP with the *Deepwater Horizon* incident in late April 2010.

BP's share price by June 25, 2010 (1 week after the congressional hearing with BP's CEO), had fallen to about 46.5% of its original value. Even in August 2013, more than 3 years later, the organization has not yet fully recovered to preincident levels and continues to expend huge sums in public relations campaigns, dealing with legal concerns related to the incident, and internal overcompensation in its bid to avoid repeat or similar incidents. Incidents are costly, and good governance should seek to avoid them.

Environmental protection, social responsibility, and health and safety at work are of significant importance to corporations, and many organizations within the oil and gas sector will be very clear and vocal in their commitment to corporate social responsibilities (CSRs). Maclagan (1998) advised that trust in organizations by all their stakeholders, including the employees, customers, and the public, is essential for their longevity and sustainable existence and growth. This has led to organizations developing systems such as audit committees, codes of ethics, and CSR-type policies. Bain and Band (1996) advised that the value of corporate governance goes beyond control in that it creates an environment of enterprise and best professional practice to extract the long-term value from a commercial enterprise.

With respect to health and safety, the Safety and Health Sustainability Task Force set up by the American Society of Safety Engineers (ASSE, 2010) has developed a safety and health sustainability (CSR) index that is built on six key elements:

1. Values and beliefs: Safety and health responsibility commitment.
2. Codes of business conduct.

3. Operational excellence: Integrated and effective safety and health management system.
4. Professional safety and health competencies.
5. Oversight and transparency: Senior leadership oversight and safety and health.
6. Transparent reporting of key safety and health performance indicators.

Since the early adoption of the ISO 9001 quality management system standard (which started as British Standard 7750), we have seen the ISO 14001 environmental management system and OHSAS 18001 health and safety management system standards. These initiatives have encouraged organizational behavior toward more self-driven compliance and certification. Such certifications have given organizations an effective, brand value proposition and marketing edge against their competitors. Such behaviors are a reflection of the appetite of organizations to invest and comply with a standard that adds value from both the internal and external perspectives.

Haefeli et al. (2005) explains in an extensive research study funded by the UK's Health and Safety Executive:

> Most organizations were concerned about potential cost implications of major incidents, but were less concerned about actual costs incurred as a result of more frequent, minor events. The majority of respondents reported that they did not know how much either accidents or work related illnesses were costing their business. Few attempts had been made to quantify the cost of health and safety failures. Limited time and resources, perceived complexity and lack of expertise were the most commonly cited barriers to conducting accident/work-related ill health cost assessments. (p. 5)

This is an important finding given the extensive research done by many safety practitioners and institutions that has linked statistically, through both cross-sectional and longitudinal studies, near-miss and minor incidents as a predictor of damage and major incidents, including fatalities.

Although most of the managers interviewed in the Haefeli et al. (2005) study were concerned about the major accidents, they did not realize that controlling or reducing minor incidents actually helps to prevent major incidents, while simultaneously improving the bottom line of the organization. Haefeli et al. (2005) advised:

> Exploration of the human impact of injury and ill health from the perspective of individual employees, their colleagues and families (e.g. psychological, social and financial effects), may be a useful angle from which to assess the impact of health and safety failures. Such information may be beneficial in assisting employers to achieve behavioral changes among staff at lower levels within organizations, as well as tapping into the moral obligations of senior managers and boards of directors. (p. 170)

In the major investigation reports of some of the most significant accidents in recent times, such as the BP Texas refinery explosion, 2005 (Baker et al., 2007; Mogford, 2005), the Piper Alpha incident in the 1988 (Kumar, 2007; Cullen, 1990), and the explosion/fire at Buncefield oil terminal in 2005 (Allars, 2007), findings clearly point out management and leadership failures in preventing such incidents.

These findings have, in fact, become a very important subject, as many of these industrial incidents, including the most recent 2010 Gulf of Mexico Macondo incident involving BP (which has been credited as perhaps the most environmentally tragic and expensive oil and gas-related incident in history), point to leadership and governance failures. Organizations that are determined to avoid such industrial disasters must ensure focused attention to leadership and governance of the organization. In so doing, the focus shifts from shareholder profit maximization to stakeholder value maximization. Leaders who understand that good safety means great business shall flourish, and those who don't will lead us down the path of repeat incidents of similar magnitudes.

The spirit of governance should be accepted and appreciated, rather than the extension of bureaucracy that corporate governance can be seen to bring. It is not adequate to ensure that the health and safety policy has been written and communicated. Rather, it is more important that stakeholders have received it, understood it, and appreciated what it actually means to their lives on and off work. Today, the focus on financial performance is shifting toward overall business performance to include performance in nontraditional focus areas such as the Health and Safety Executive (HSE), CSR, resources management, and environmental performance, including greenhouse gas production and management. Therefore, as Zukis et al. (2010, p. 1) pointed out, stakeholder perspectives on organizational performance now reside in the following areas:

1. Governance and internal controls
2. Operations and risk management
3. Organizational change
4. Performance improvement

The subject of corporate governance is vast and touches upon many aspects of an organization and its stakeholders. It has become an expectation for organizations to control their operations and to show control, especially with aspects that touch upon the people, environment, and general public.

This discourse provides a great introduction into operational excellence management systems (OEMSs). Corporate governance is a critical part of the overall contributors to organizational quests for OEMS. When organizational governance fails to execute its required roles and responsibilities, it should be held accountable in similar fashion as organizational leadership. The focus on financial maximization vs. value maximization has to be front and center in the agendas of all corporate governance personnel.

References

Allars, K. (UK-HSE). (2007). Buncefield, the story so far. Case study presentation delivered at the Platts: Creating value in oil storage, Budapest, Hungary, November 26–27. Retrieved March 6, 2013.

American Society of Safety Engineers (ASSE). (2010). Safety and health sustainability index taskforce. Working Paper. Retrieved January 4, 2013, from http://centershs.org/archive/documents/TC_Safety_Health_Sustainability.pdf.

Bain, N., and Band, D. (1996). *Winning ways through corporate governance*. 1st ed. Macmillan Business.

Baker, J., III, Bowman, F.L., Glen, E., Gorton, S., Hundershot, D., Levison, N., Priest, S., Rosentel, T.P., Wiegmann, D., and Wilson, D. (2007). The report of the B.P. U.S. refineries independent safety review panel. Retrieved August 22, 2013, from http:///www.csb.gov.

Bani Hashem, W.S.G. (2011). Safety leadership case study—An in-depth analysis of the congressional hearing with Tony Hayward, BP ex-CEO on the *Deepwater Horizon* BP oil disaster, 2009 in the Gulf of Mexico, USA. Working Research Paper 01/11. Presented at 9th Annual OHS Congress, Dubai, UAE, February 5–10.

British Safety Council (BSC). (2005). International diploma in occupational safety and health, module A. British Safety Council.

Cadbury Committee Report. (1992). *The Financial Aspects of Corporate Governance*. Gee and Co.

Chew, D.H., and Gillan, S.L. (2005). *Corporate governance at the cross-roads—A book of readings*. McGraw-Hill/IRWIN Services in Finance, Insurance and Real Estate.

Clutterbuck, D., and Waine, P. (1993). *The independent board director—Selecting and using the best non-executive directors to benefit your business*. 1st ed. McGraw-Hill.

Collins, J. (2001). *Good to great*. 1st ed. HarperCollins.

Cullen, W.D. (1990). *The public enquiry into the Piper-Alpha disaster*. HMSO.

Det Norske Veritas (DNV). (1996). *Loss control management, modern safety management training program. Module on the causes, effects and control of loss*. Det Norske Veritas.

Dunlop, A. (1998). *Corporate governance and control, business skills series*. 1st ed. Chartered Institute of Management Accounts (CIMA).

Goodpaster, K.E., and Matthews, J.B. (1982). Can a corporation have a conscience? *Harvard Business Review*, January.

Haefeli, K., Haslam, C., and Haslam, R. (2005). Perceptions of the cost implications of health and safety failures. A research report prepared for the Health and Safety Executive. Research Report 403. Institute of Work, Health and Organizations and the Health and Safety Ergonomics Unit, Health and Safety Executive (HSE) Books.

Handy, C. (2002). What's a business for? *Harvard Business Review*, December 2002.

Health and Safety Executive (HSE). (1996). *Cost of accidents at work guideline*. HSG-96. HSE Books. Retrieved August 2, 2013, from http://www.lse.co.uk/ShareChart.asp?sharechart=BP.&share=bp.

Health and Safety Executive (HSE). (2001). *A guide to measuring health and safety performance*. HSE Books.

Institute of Chartered Accountants (IoCA). (1999). Internal control—Guidance for directors on the combined code. Turnbull Report. Institute of Chartered Accountants.

Kendall, N., and Kendall, A. (1998). *Real-world corporate governance—A programme for profit-enhancing stewardship*. 1st ed. Pitman Publishing.

King, M.E. (2009). King code of governance for South Africa 2009. Institute of Directors, South Africa.

Leavey, B., and McKiernan, P. (2009). *Strategic leadership—Governance and renewal*. 1st ed. Palgrave Macmillan.

Maclagan, P. (1998). *Management and morality. A development perspective*. 1st ed. Sage.

McGregor, L. (2000). *The human face of corporate governance*. 1st ed. Palgrave.

Mogford, J. (2005). Fatal accident investigation report. Isomerization unit explosion. Final Report. BP Texas Refinery, Texas City.

PricewaterhouseCoopers (PWC). (2005). Corporate governance. Governance, risk and compliance series—Connected thinking, global best practices. PricewaterhouseCoopers.

Zukis, B., Quan, J., Bala, S., and Minakawa, X. (2010). Policies and procedures for operational effectiveness—Enabling a framework for continuous improvement. Best practices for making policies and procedures a platform for operational effectiveness. PricewaterhouseCoopers.

2

High-Reliability Organizations (HROs) and Operational Excellence

2.1 Introduction

As discussed earlier, high-reliability organizations (HROs) are generally resilient organizations that have effective systems with redundancy built in to them. Process safety management has helped move many organizations along the continuum of improved performance, higher reliability, and operational excellence. However, there are probably very few process safety management (PSM) elements that are as impactful on process safety reliability as the management of change elements and its requirements. We believe that the development and improvement of this process can be explained through the building of high-reliability organizations.

2.2 Defining High-Reliability Organizations

The Health and Safety Laboratory (HSL, 2011) provided one of the most comprehensive studies defining high risk and high reliability. In this work, the definitions start from the context of the two (competing) prominent schools of thought that seek to explain accidents in complex, high-hazard organizations:

1. Normal accident theory (NAT)
2. High-reliability organization theory (HROT)

According to NAT, the definition is very straightforward and depicts the tight coupling of various aspects and system components (e.g., people, equipment, procedures). Due to the complex relationships and interdependencies of these tightly coupled and often highly automated systems, the timing of tasks does not even allow for human intervention.

Perrow (1984), as cited by the Health and Safety Laboratory (HSL, 2011), explains that when a failure occurs in one part of the system, it quickly spreads to another part of the system and results in a massive failure. Interestingly, Perrow classified petroleum and petrochemical plants, such as refineries, as lower risk than military systems and aircrafts, etc. This theory was highly criticized mainly for failing to be consistent in accurately capturing and differentiating between the design features of systems in these industries. Others have identified the weakness in the definition of the theory itself through its coupling and complex terminologies. Its currency is also of little value to practitioners, as it fails to advise and suggest how accidents can be reduced. In essence, the NAT describes the consequences of the human-machine interface actions, where the source of failure is more physical or technological.

In contrast, the HROT can be described as more in response to uncertainty, complexity, and risks. Here, the focus is more on the behavioral and sociophysical aspects (see http://high-reliability.org). The sociophysical dimension is created from the tight coupling between the human being and the machine/physical processes. The definition of HROT addresses the criticisms of NAT. This theory advocates that accidents in high-hazard, complex systems are not inevitable because the processes in place enable these organizations to effectively prevent incidents and contain catastrophic errors, thereby maintaining a consistent record of safe operations.

In fact, HROT researchers maintain a positive view with regard to the nature of accidents in complex systems by arguing that organizations can become more reliable by creating or engineering a positive safety culture and reinforcing safety-related behaviors and attitudes. What is very interesting here is that HROT researchers maintain that such organizations are not error-free as much as they are preoccupied with failure and prevention of that failure, and how to deal with failing systems. Most significantly, such organizations exhibit strong learning orientation, prioritization of safety over other goals, continual training and development, and an emphasis on checks and maintaining the safety performance.

To this end, they also explain that HROT perspectives have much in common with resilience engineering, which are systems employed extensively in the aviation, petrochemical, and nuclear industries. However, the HROT has also had its fair share of criticism for ignoring the broader social and environmental contexts to learn from errors. Examples quoted include the (corporate) political implications of errors that may impact on the extent to which errors can be openly reported. It is important to understand the actual characteristics of high-reliability organizations (HRO), which are summarized in the Table 2.1.

The Organization for Economic Cooperation and Development (OECD, 2012) offers a definition of HROs as one "that produces product relatively error-free over a long period of time" (p. 10). Two key attributes are described, including having *a chronic sense of unease* and therefore lacking the sense of

TABLE 2.1
Attributes of a HRO

No.	Characteristic	Implication
1	Dynamic leadership shift	• Hierarchical decision making during routine periods. • Clear responsibilities during emergencies. • The organization migrates to a structure in which it leverages members within the organization who have the expertise.
2	Systematic intervention	• They manage by exception, and thus managers focus on strategic and tactical decisions and seldomly interfere with operational issues that are delegated and covered by clear processes.
3	Learning organization	• Climate of continuous training and learning.
4	Multicommunication	• Several channels are used to communicate safety-critical information—timely communication of information during normal and emergency situations.
5	Redundancy	• Built-in redundancy and the provision of backup systems in case of a failure.

complacency. They, as such, can be described as believing that an incident can happen at any time, even if no incidents have taken place for a very long time. The second important attribute is to "make strong responses to weak signals" (p. 10), and therefore to set a low threshold for intervention. This generally means that they will go to the extent of shutting down operations to investigate more frequently, which may mean financial losses. They see this as an essential risk control measure to prevent a potentially larger loss.

2.3 Developing High-Reliability Organizations

HSL (2011) noted that resilience in HRO can be engineered by incorporating the following characteristics:

1. A just culture that promotes transparency in reporting of incidents and improvements, with a great balance between supporting the reporting culture and tolerating unacceptable behaviors
2. Management commitment that balances the pressures of production with safety and management behavior/allocation of resources
3. Increased flexibility through supportive systems and empowerment
4. Learning culture in which information is shared, regular effective training is undertaken, and there is development of critical safety information

5. Preparedness through proactive safety management systems
6. Opacity/awareness through organizational collection and analysis of information that enables the organization to identify hazards and risks early and deal with prevention
7. Resources—in the form of competent staff, systems, technology, and additional resources to help prevent incidents and deal with them when they happen

This is consistent with Ghanem Al Hashmi's (2011) ADIPEC 2 reference model of safety culture in which awareness borne from information sharing and training, as well as autonomy and management support, is an indicator of a safety culture. More could be discussed with respect to safety culture, because it seems that there is little difference between a strong safety culture within an organization and high reliability and performance. They share many similar attributes; however, the HRO definition is more encompassing.

Al Hajri (2008), in a case study of Al Jubail Petrochemical Company (KEMYA), explained that management commitment and leadership are the key drivers for the KEMYA behavior-based safety (BBS) proactive program. In this instance, the safety performance, management performance to process safety management (PSM), and the operational integrity management system leading to its safety excellence program (SEP) are all linked. Operational excellence is a function of high reliability and will be discussed later in this chapter.

The aviation industry and industries with process, manufacturing, and hydrocarbons are inherently risky, and are characterized by evolving risk profiles from changing conditions and environments. Maintaining high reliability becomes a fundamental cornerstone of the very viability and sustainability of businesses involved in these industries. HROs generally develop their strengths through the actions of individuals who share highly aware and safe attitudes within the organizations, which in turn, over time, creates an organizational culture that can be described as a *high-reliability culture.*

There are four very fundamental organizational characteristics that help HROs control the number of incidents that occur. According to High Reliability Organizing (2014), these include the following:

1. A prioritization of safety and shared performance goals throughout the organization
2. An organizational culture of reliability as described above
3. A learning organization that uses higher orders of learning to continually improve
4. A strategy of redundancy beyond technology

There are inherent difficulties with defining HROs mainly because incidents occur all the time in organizations as near hits or—as commonly referred to in the industry—near misses. The majority do not eventually become an incident; i.e., these incidents do not result in loss of some kind. The question of defining HROs quantitatively raises this issue with many of the definitions, as an industry benchmark is required in aerospace, aviation, oil and gas, manufacturing, and other industries. So statistically, if a near-accident-free performance is achieved, then perhaps that particular organization can be described as an HRO.

The above also means that statistically, reliability can be calculated, but this would be a function of uncertainty—a fundamental issue of HROT. If and when HROs are defined statistically, there is a degree of engineering/technical/mathematical accuracy, but the factors related to organizational and social matters bring about greater uncertainty to some extent (Marais et al., 2006). This potentially explains why at the heart of the OECD PSM/CG model, there is both leadership and (organizational) culture.

Weick and Sutcliffe (2001), as cited by Hopkins (2007), discussed the five characteristics of HROs designed to produce a collective state of mindfulness. These include the following:

1. Preoccupation with failure
2. Reluctance to simplify interpretations
3. Commitment to resilience
4. Sensitivity of operations
5. Deference to experience with the encouragement of a fluid decision-making system

While Hopkins (2007) supported their views, he also advised that the challenge in defining HROs thus lies in the very fact that a detailed inquiry looking at these five areas would be required, and this was highly dependent on the industry within a context of time.

Rooksby (2010, as cited by Marais et al., 2006) explained that managers in HROs work closely with their subordinates about their work actions, rather than just focusing on figures related to bottom-line performance. Therefore, in a way, it is enhancing performance through creating learning organizations. It is worthy to note that given the above, some studies have found that the HRO researchers have oversimplified the complexity and difficulties that engineers and scientists face, and have suggested an alternative systems approach to safety, which tries to overcome the limitations of both the NAT and HRO theories (Marais et al., 2006).

The human element remains very critical, as noted by Bridges (2010). He argued that weak management systems compounded with human errors cause incidents. Therefore, human factors must be understood very well in the prevention of incidents. Occupational health performance standards, training and competency, task design, manpower rationalization and time motion studies to determine the safe managing levels, task-human-system interaction, etc., all help strengthen PSM in businesses.

In a recent and dynamic study of an unnamed oil refinery in the UK that was actively working toward higher levels of reliability and safety, in response to the Texas City explosion in 2005, four fundamental areas or themes emerged from both one-to-one and focus group interviews. Themes identified by Lekka and Sugden (2011) that contributed to higher reliability and safety included the following:

1. Training and technical competence
2. Hazard identification and awareness
3. Learning orientation
4. Strong management commitment to safety

However, among the conclusions of the study, management commitment and high levels of management visibility may be among the challenges faced by organizations striving to implement high-reliability practices.

Finally, since HROs have a migratory decision-making processes that allows those who are closer to an incident to react to prevent escalation or otherwise, the *empowerment within the hierarchy* to those who would be better informed (or more knowledgeable), and the leadership model within these organizations, would have to be flexible and very empowering. Furthermore, under normal conditions, leaders would be required to be engaged and have a good grasp or appreciation of the risks and challenges faced within the day-to-day operations of the organization.

Clearly, therefore, leadership from the top must be very trusting and capable, but would, within the context of operational integrity, need to expect inherently highly reliable operations. This is particularly the case with operations such as oil and gas, power and utilities, aviation, etc. This suggests that operations are managed within a reliable, integrated management system, where the changes in response can be fast enough to deal with any rapidly occurring developments. This also requires that managers at all levels, starting from the top of the organization, are trained to manage and respond. This can only be achieved through structured systems, training, and drills. It can be argued, therefore, that a culture of high reliability must be driven by the executive management team and the board of directors, who would set the tone as an expectation.

2.4 Linkage between Operational Excellence Systems and High-Reliability Systems

The development of operational excellence management systems requires strong integration among leadership, environmental health and safety (EHS), quality management, and other critical elements, such as operating processes, control of work, procedures, and asset/mechanical integrity. Having effective systems in place helps prevent incidents by ensuring integrity and preventing the failure of safety barriers, which are designed for incident prevention.

Most high-risk industries are highly regulated. However, even within highly regulated industries, management systems function to facilitate smooth and incident-free operations. These processes and procedures are built on the foundations of quality management systems. It is the integration of leadership, quality management systems, and fit-for-services EHS management systems with supporting processes and procedures that creates an operational excellence management system.

Finally, in some recent research by Al Hashmi (2014), when comparing the oil and gas sector to the non-oil and gas sector, a statistically significant difference in perspective was found to exist. Al Hashmi found the degree of innovation in the oil and gas sector remains higher, and much of this is attributed to the fact that many of the processes, procedures, and systems within this industry have been driven by very large oil and gas companies. By comparison, the marine, aviation, and manufacturing industries have had a greater influence placed upon them by the international bodies, associations, and authorities that have clearly established operating envelopes and system requirements; i.e., they have a more prescribed system of working rather than a system developed more internally and exclusively by that particular organization.

References

Al Hajri, M.Q. (2008). BBS leading to safety excellence. ASSE-MEC-0208-38. American Society of Safety Engineers, Middle East Chapter Conference and Exhibition, Bahrain.

Al Hashmi, W.G. (2012). Assessment of safety culture through perception studies—Using quantitative methods in management research—Case study from the Emirates National Oil Company (Ltd) LLC Group of Companies. Working Research Paper 02/12. Presented at ADIPEC, Abu Dhabi, UAE, June.

Al Hashmi, W.G. (2014). EHS leadership and governance in high risk organizations: Exploring perspectives from the GCC. Doctoral thesis (working stage).

Bridges, W. (2010). Human factors elements missing from PSM. ASSE-MEC-2010-48. American Society of Safety Engineers, Middle East Chapter Conference and Exhibition, Bahrain, p. 392.

Health and Safety Laboratory (HSL). (2011). *High reliability organizations—A review of the Literature*. Health and Safety Executive Research Report RR899. HSE Books, 2011.

High Reliability Organizing. (2014). Managing the unexpected. Retrieved April 2, 2014, from http://high-reliability.org/pages/High-Reliability-Organizations.

Hopkins, A. (2002). Safety culture, mindfulness and safe behaviour: Converging ideas. National Research Centre for OHS Regulation, Australian National University, December.

Lekka, C., and Sugden, C. (2011). The successes and challenges of implementing high reliability principles: A case study of a UK oil refinery. *Process Safety and Environmental Protection*, 89(6), 443–451.

Marais, K., Saleh, J.H., and Leveson, N.G. (2006). Archetypes for organizational safety. *Safety Science*, 44(7), 563–582.

Organization for Economic Cooperation and Development (OECD). (2012). Corporate governance for process safety—Draft guidance for senior leaders in high hazard industries. Draft. Environment Directorate, Joint Meeting of the Chemicals Committee and the Working Party on Chemicals, Pesticides and Biotechnology.

3

Types of Management Systems

3.1 Introduction to Management Systems

Perhaps every organization today operates with some form of a management system. Management systems vary from the extremely sophisticated, with well-crafted standards and policies guiding the behaviors and actions of all workers, to the informal organization, with loosely held work practices and behaviors adopted by the leader of the organization. Regardless of the type of management system, it is an essential requirement to manage the performance of an organization. Some of the more clearly defined and developed management systems are adopted and used by today's leading organizations. Among them are the following:

- Operations integrity management system (OIMS)—ExxonMobil
- Operations management system (OMS)—BP
- Operational excellence management system (OEMS)—Suncor Energy
- Operational excellence management system (OEMS)—Chevron

The authors define a management system as a series of policies, standards, procedures, and work practices that govern *people*, *processes*, and *facilities* management within an organization. The discipline applied in conforming to these policies, standards, procedures, and work practices determines the ultimate business performance of the organization. Some may argue that management systems are required only for large organizations; however, the authors suggest that regardless of the scale and scope of operations, businesses do better with a well-crafted management system. Executing a management system across a small organization is significantly less difficult than doing so across a larger organization, and requires a disciplined approach.

In a recent discussion with a close relative—a budding entrepreneur who gave up a promising career in a large multinational organization to operate her small retail coffee shop—she provided an analogy that explained her experiences. She advised that her experiences were similar to cooking a quality meal in a kitchen. In her view, the requirements for providing the same quality meal were independent of the number of people being fed. She advised the common requirements for providing a quality meal include a high standard of preparation, clean utensils, an organized table setting, and a supervised meal preparation. She advised that the management system requirements for her small coffee shop were essentially no different from those of her prior employer. The only difference, in her view, was that she now stewarded all of the requirements of the management system, whereas in her prior employed situation, some requirements were stewarded by various departments and functions in the organization.

She advised that the principles of business in her small coffee shop were essentially the same as those of the large multinational corporation, among which common practices such as the following surfaced:

1. *People management*: Hiring, training, competency assurance, supervision, and remunerating.
2. *Processes and systems*: Procedures, operations management, supply chain management, accounting, marketing, and housekeeping.
3. *Facilities and technology*: Siting, quality assurance, mechanical integrity, preventive maintenance, and contractor management.

Developing and maintaining a management system provides any organization a roadmap for exceptional performance and success. Indeed, for larger multinational corporations, a management system is a critical component for sustainable growth and performance management.

In this book, the authors discuss a series of fundamentals for creating an operationally disciplined and excellent management system. The authors explore each fundamental in detail and provide the supporting training and workshop materials that are essential for integrating these fundamentals into the business processes of the organization. The authors also incorporate sound organizational and business practices with personnel and process safety management principles to craft a compelling argument for creating and sustaining an operationally disciplined and excellent management system for exceptional business performance.

The authors identify seven fundamentals for an operationally disciplined and excellent management system:

1. Leadership commitment and motivation
2. Defining the applicable elements for an operational excellence management system
3. Establishing the baseline

4. Following the requirements of the plan–do–check–act model
5. Auditing for conformance and compliance
6. Closing the gaps—operational discipline
7. Sharing learning and continuously improving

This compilation of practical principles for each fundamental used, to varying degrees, in the workplace provides excellent opportunities for organizations to excel in today's business environment.

The scalable workshop materials included in this book provide organizations an opportunity to modify the content of the materials to meet their specific needs for training and educating workers at all levels of the organization. The content of this book also leverages prior work done by Lutchman in project execution (Lutchman, 2010), safety management (Lutchman et al., 2012), and process safety management (Lutchman et al., 2013) for continuous improvements in businesses processes and practices.

In this chapter, the authors provide an introduction to management systems. Readers are provided with a sense of the evolution of organized management systems in business. The authors also seek to identify the underlying principles and elements of the management system so that readers are made aware of the continual evolution of the organization. In addition, they will understand the quest for business performance improvements in a business environment characterized by intense competition, dwindling resources, demands for greater social responsibility, and rapid transfer of knowledge and information.

3.2 Definition of a Management System

Taken separately, *management* refers to the day-to-day tactical strategies and control of actions to achieve business targets and goals. A *system* is defined as a set of organized and interconnected processes used to achieve a defined business purpose. In its very basic form, therefore, *a management system* can be defined as the application of a set of organized and interconnected processes during the employ of day-to-day tactical strategies and control of actions to achieve business targets and goals.

The authors define a management system as a series of standardized requirements developed in elements, policies, standards, and procedures designed to direct the behaviors and work performance of all levels of the organization to achieve the strategic and tactical goals of the organization, and tactical as well as overall business performance. A management system embraces the following:

1. Leadership commitment
2. Clearly defined elements and their requirements

3. Establishing a baseline
4. Following a plan–do–check–act model for managing work
5. Performance management and review
6. Maintaining operational discipline
7. Continuous improvement

The teachings on management and system of profound knowledge of W. Edwards Deming, one of the founding fathers of quality management, are still applicable today in that he believed that true, profound leadership is not something inherent, but rather, it can be created and cultivated in individuals by building skills in a number of specific areas. Leaders can be found at all organizational levels and can rally people to a cause because their skills and abilities match a particular set of follower needs and situational circumstances. These are competencies that can be learned and practiced by anyone with the motivation to step forward (Schultz, 2013).

Ownership of the system resides in the hands of senior leaders. Execution of the requirements of each element is achieved at the front line through the guidance and actions of the management from direct supervision and consistent interpretations of the requirements of each element, policy, standard, and procedure.

3.3 Types of Management Systems

Among the management systems that have influenced our business performance and continue to do so in a meaningful way today are the following:

- Total quality management (TQM)
- International Organization for Standardization (ISO) 9000 and family
- Process safety management (PSM)
- Operational excellence management system (OEMS) and its variations

The authors review each of these in turn to provide users with a better understanding of their attributes.

3.3.1 Total Quality Management (TQM)

Generation X leaders may remember the introduction of total quality management (TQM) principles in the 1950s. Quality management is a result of the work of quality gurus: the American gurus featured in the 1950s, Joseph Juran, W. Edwards Deming, and Armand Feigenbum; the Japanese quality gurus who developed and extended the early American quality ideas and models,

Kaoru Ishikawa, Genichi Taguchi, and Shigeo Shingo; and the 1970 and 1980s American gurus, notably Philip Crosby and Tom Peters, who further extended the quality management concepts after the Japanese successes, including the development and use of quality tools such as Deming's plan–do–check–act (PDCA) cycle, conveying that quality improvement is a continuous process.

The early work of Walter Shewhart, Harry Deming, Harold Dodge, and Walter Edward Romig constitutes much of what today comprises the theory of statistical process control (SPC). After World War II, Japan decided to make quality improvement a national imperative as part of rebuilding its economy, and sought the help of Shewhart, Deming, Juran, and Feigenbaum, among others. In 1969 the first international conference on quality control, sponsored by Japan, America, and Europe, was held in Tokyo. In a paper by Feigenbaum, the term *total quality* was used for the first time, and referred to wider issues such as planning, organization, and management responsibility. Ishikawa presented a paper explaining how "total quality control" in Japan was different. It meant companywide quality control and described how all employees, from top management to the workers, must study and participate in quality control. Companywide quality management was common in Japanese companies by the late 1970s. In fact, during this period, Japan's imports into the United States and Europe increased significantly due to their higher quality and better costs compared to the Western counterparts (Shirshendu and Prasun, 2011).

Among the many early adopters of TQM principles in its business practices was the Japanese car company Toyota. The quality revolution in the West was slow to follow, and did not begin until the early 1980s, when companies introduced their own quality programs and initiatives to counter the Japanese success. Total quality management became the center of these drives in most cases. TQM took hold in the 1980s, helping organizations to become more customer focused, with products and services that satisfy their needs. TQM is both a philosophy and a set of guiding principles for managing an organization to the benefit of its stakeholders. TQM, put simply, is the mutual cooperation of everyone in an organization, including customers and suppliers, and their integration with key business processes to produce value for money products and services that exceeds the needs and expectations of customers. The process of improvement also concentrates on the elimination of waste and non-value-added activity (Dale et al., 2007; Shirshendu and Prasun, 2011). Hashmi (2010) advised that TQM defines the culture, attitude, and organization of a company. In such organizations, quality is demonstrated in all aspects of the company's operations, such that work is done *right at all times and rework defects and wastage are eliminated*. The fundamental principles of TQM include the following:

1. Customer focused
2. Total employee involvement

3. Process centered
4. Integrated system
5. Strategic and systematic approach
6. Continual improvement
7. Fact-based decision making
8. Communications

Throughout the 1980s, U.S.-based manufacturing operations continued to lose ground to constantly improving foreign competition, partially because Japanese firms believed in helping each other and sharing best practices with their countrymen. Then, in 1987, the U.S. government introduced the Malcolm Baldrige National Quality Award (MBNQA). This award, presented annually by the president, is designed to provide an operational definition of business excellence. Two key aspects of the Baldrige Award are the promotion of best practice sharing and the establishment of a benchmark for quality systems that focused on customer satisfaction as a primary driver of business design and execution (Folaron, 2003).

There are a number of quality management tools and techniques that are a fundamental part of an organization's road to TQM. These include a variety of basic quality control tools, such as the PDCA cycle, flowcharts, histograms, graphs, Pareto analysis, cause-and-effect diagrams, brainstorming to quality function deployment, design of experiments, failure mode and effects analysis (FMEA), statistical process control (SPC), Six Sigma, Lean management, and business process reengineering (Dale et al., 2007).

3.3.1.1 Six Sigma

Six Sigma is a quality management initiative that is a data-driven, methodological approach to eliminating defects with the aim of reaching six standard deviations from the desired target of quality. Six standard deviations means 3.4 defects per million. The Motorola Company introduced the Six Sigma approach to quality and won the Malcolm Baldrige National Award in 1988. It recognized the need for focused quality improvement, and the award simply confirmed it had an approach and deployment of metric-based, customer-focused quality that would lead to the current Six Sigma methodology. The company's approach to continuous improvement was based on a comparison of process performance and product specification, and aggressive efforts to drive down defects. As a result of winning the Baldrige Award in 1988, Motorola was compelled to share its quality practices with others. The team that developed Six Sigma became the core of the "Motorola University," teaching these skills to other organizations globally, and establishing internal universities in other organizations. That experience has become a new model for high-profile internal training programs worldwide.

3.3.1.2 Lean Management

Lean management is a process of maximizing customer value while reducing waste. Any activity or process that consumes resources and adds cost or time without creating value becomes the target for elimination. Lean Six Sigma and other similar programs have helped organizations find breakthrough solutions to their most costly process problems. The Lean Six Sigma approach offers a great model for practical, everyday problem solving.

3.3.1.3 Business Process Reengineering (BPR)

Business process reengineering (BPR) enables an organization to make a radical and revolutionary examination of the way it operates and how work is done. BPR emphasizes structural process redesign, process reengineering, and fundamental business rethinking by ignoring the status quo, and it looks for savings over a shorter period of time (Dale et al., 2007).

TQM is credited as a driver of continuous improvements and innovations in organizations. More importantly, TQM has driven business processes to higher levels of efficiency and generated competitive advantages, allowing some organizations to reduce costs and improve margins while at the same time delivering quality products to their customers. TQM encourages organizations to innovate within product lines and business processes to become more efficient and produce exceptional business performance.

Honarpour et al. (2012) pointed to defined relationships among innovation, total quality management, and knowledge management. They also cited studies that confirmed positive association between TQM and innovation. They suggested that TQM also influences innovation in the following three ways:

1. Organizational practices and responses to market orientation and customer focus
2. Continuous improvement strategies
3. Teamwork, employee empowerment, and people management strategies that promote autonomy and knowledge sharing among employees

Honarpour et al. (2012) conceptualized the model for innovation and its relationship with TQM and knowledge management as shown in Figure 3.1.

According to Khasawneh et al. (2012), today TQM is "considered by organizations to be an effective way to bring about radical changes in philosophy and style in the way work is done in order to achieve the highest levels of quality and to be used as a bridge to higher customer satisfaction and retention" (p. 2). Indeed, one can appreciate the value of customer retention since returning customers build loyalty and contribute to the overall organizational performance.

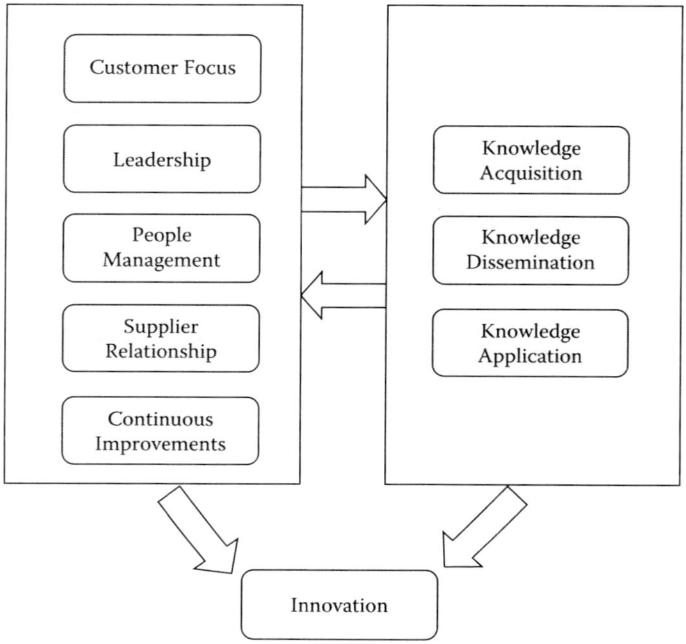

FIGURE 3.1
Conceptual model for innovation derived from TQM and knowledge management. (From Honarpour, A. et al., *Journal of Technology Management and Innovation*, 7(3), 22–31, 2012, retrieved February 21, 2013, from EBSCOhost database.)

Khasawneh et al. (2012), in a study of the Jordanian industrial sector, saw TQM as an approach for remaining competitive in today's global landscape. They recommended the following for governmental institutions:

- Continue to increase interest in TQM principles or variants of them.
- Specify strategic and tactical goals of TQM, and support them with appropriate policies and procedures. Identify and leverage competitive advantages for organizations to succeed in today's global, competitive business environment.
- Plan appropriately as to include specific quality goals set by management.
- Adopt a scientific approach to improving quality.
- Ensure quality requirements are entrenched in organizational control and structure.
- Promote worker participation and empowerment for preparing plans to develop quality.
- Work toward establishing annual training plans to develop worker skills and abilities.

In summary, TQM is a method by which management and employees can become involved in the continuous improvement of the production of goods and services. It is a combination of quality and management tools aimed at increasing business and reducing losses due to wasteful practices.

There is no single route leading to success in TQM, as different management styles and corporate cultures will need to take different paths in organizations. The introduction and subsequent development of TQM must be led by the company's leaders, and it must be accepted from the onset that it will be liable to setbacks owing mainly to resistance to change. The Shitsuke principle from the Kaizen 5S framework, meaning "sustaining the discipline," is an important and often overlooked aspect of TQM.

It can arise from a variety of factors, but by and large, it sets in after the passage of time. This difficulty in sustaining a discipline can occur due to complacency, culture, changes, and cost, among other things. Systematic and long-term planning, including proper resource utilization, must be a major part of the implementation of the TQM process. In order to maintain progress, senior management must proactively monitor the process using key performance indicators (KPIs). Maintaining a clear focus on customer wants while continuously improving the processes that provide the product or service will be the only way for organizations to remain competitive. Top management and leadership must learn and be committed to have an effective and sustainable TQM philosophy.

3.3.2 ISO 9000

The desire for consistency in the definition of quality led to the development of industrial standardization organizations, beginning in Great Britain in 1901. By 1930, most of the world's industrialized nations had similar organizations. In 1987, the Geneva-based International Organization for Standardization (ISO) introduced a series of quality standards that were adopted by most of the industrialized world to serve as single, global standards. These global standards were designed to promote uniformity between countries that had their own definition of quality specifications. ISO is the world's largest agency for voluntary international standards and comprises the national standards bodies of more than 160 countries (Folaron, 2003). The ISO 9000 series of standards was thought to support the delivery of quality management systems and was designed to help organizations meet agreed quality and certification requirements for products offered to customers and other stakeholders.

ISO management systems provide a model to follow when setting up and operating a management system, as ISO standards are based on the principle of continuous improvement, and therefore organizations' policies can be continually reviewed and improved, which aids in making industries more efficient and effective. It comprises a number of standards that specify the requirements for the documentation, implementation, and maintenance of a quality system. By implementing voluntary ISO standards, companies make a proactive commitment to the principles they stand for: quality, transparency, accountability, and safety. Since the early 1990s, ISO 9000 standards and management principles have led the global drive for quality practice in business. According to Majstorović and Marinković (2012), these ISO management principles include the following:

1. Customer-focused organization
2. Leadership
3. Involvement of people
4. Process approach
5. System approach to management
6. Continual improvement
7. Factual approach to decision making
8. Mutually beneficial supplier relationships

Successive revisions of ISO 9000 led to the 2008 set of ISO 9001 standards that are now being adopted by organizations for global standardization of product quality and business processes.

While the ISO 9000 and ISO 14000 families are among the widely known international standards, there are other ISO international standards that provide practical tools for tackling many of today's global challenges that impact our lives. These standards range from managing scarce global water resources to food quality and safety to tackling climate change by implementing standards for measuring greenhouse gas emissions and promoting energy efficiency to standards for computers, cars, agriculture, healthcare, services, and even risk management. Organizations benefit from implementing these standards without having to be certified in them. Audits are a vital part of ISO's management system approach, as they enable the company or organization to check the extent to which its achievements meet the objectives. External audits also play a role in showing conformity to the standard. Some of these standards, as described on the American Society for Quality website (http://asq.org/learn-about-quality/iso-9000/overview/other-standards.html), are summarized below.

3.3.2.1 ISO 14001

The ISO 14000 series specifies requirements for establishing an environmental policy, determining environmental impacts of products or services, planning environmental objectives, and implementing programs to meet objectives, corrective actions, and management reviews. It provides practical tools for organizations trying to control their environmental impact and constantly improving their performance. It specifies requirements for an environmental management system to enable an organization to develop and implement a policy and objectives that take legal and other requirements into account. It applies to those environmental aspects that the organization identifies as within its control and ability to influence.

3.3.2.2 ISO 16949

This is a technical specification that, in conjunction with ISO 9001, defines quality management system requirements for the design, development, production, and when relevant, installation and service of automotive-related products. It was developed by the International Automotive Task Force and aligns American, German, French, and Italian automotive quality standards within the global automotive industry

3.3.2.3 ISO 19000

Many of ISO's 19000 standards can help businesses and organizations all over the world make progress in the three pillars of sustainable development: the environment, economy, and society.

3.3.2.4 ISO 19011

ISO 19011 provides guidance on auditing management systems, with details on the principles of auditing, managing an audit program, and conducting management system audits. It also includes guidance on evaluating the competence of individuals involved in the audit process, including the person managing the audit program, auditors, and audit teams.

3.3.2.5 ISO 22000

The food industry uses the ISO 22000 series, which entails the specific requirements for food safety management systems, interactive communication, system management, prerequisite programs, and Hazard Analysis Critical Control Point (HACCP) principles.

3.3.2.6 ISO 26000

ISO 26000:2010 provides guidance on how businesses and organizations can operate in a socially responsible way. This means acting in an ethical and transparent way that contributes to the health and welfare of society. It helps clarify what social responsibility is, helps businesses and organizations translate principles into effective actions, and globally shares best practices relating to social responsibility.

3.3.2.7 ISO 27001

The information security management standard helps keep information assets secure and helps manage security (financial information, intellectual property).

3.3.2.8 *ISO 31000*

Risks affecting organizations can have consequences in terms of economic performance and professional reputation, as well as environmental, safety, and societal outcomes. Therefore, managing risks effectively helps organizations to perform well in an environment full of uncertainty. ISO 31000:2009, *Risk Management—Principles and Guidelines*, provides principles, a framework, and a process for managing risk. Using ISO 31000 can help organizations increase the likelihood of achieving objectives, improve the identification of opportunities and threats, and effectively allocate and use resources for risk treatment.

3.3.2.9 *ISO 50001*

This series provides a framework of requirements for organizations to develop a policy for more efficient use of energy, fix targets and objectives to meet the policy, use data to better understand and make decisions about energy use, measure the results, review how well the policy works, and continually improve energy management.

3.3.3 Process Safety Management (PSM)

Interest in PSM has been fueled by the continuous onslaught of catastrophic events faced by businesses on an ongoing basis. According to the Occupational Safety and Health Administration (OSHA, 2000), historical incidents leading to a more rigorous management system for addressing workplace hazards included the following:

- 1984 Union Carbide, Bhopal, India, incident that resulted in >2,000 fatalities
- 1989 Phillips Petroleum Company, Pasadena, Texas, incident that resulted in 23 fatalities and 132 injuries
- 1990 BASF, Cincinnati, Ohio, incident that resulted in 2 fatalities
- 1991 IMC, Sterlington, Louisiana, incident that resulted in 8 fatalities and 128 injuries

Introduced in 1990–1991 in the United States by OSHA, PSM and the 14 identified PSM elements brought about tremendous improvements in the way hazards are managed in the workplace (OSHA, 2000).

PSM is the application of management systems designed to identify, understand, and control process hazards to prevent process-related injuries and incidents (Lutchman et al., 2013). Patel (2012) defined PSM as the "application of management systems and controls (standards, procedures, programs, audits, evaluations) to a process in a way that *process hazards are identified, understood and controlled*" (p. 6). Patel suggested further that success in PSM requires "a sustained effort by leadership at all levels" (p. 6) of the organization and, for global organizations, a single global PSM management system may be best.

According to Lutchman et al. (2013), the principal concerns of PSM are therefore as follows:

1. *Process issues*: Concerns such as fires, explosions, toxic gases, and unintended releases and how they affect the workers in their use, storage, and disposal.
2. *Safety*: Concerns around how the company uses or implements safety regulations and any program to reduce incidents and injuries in the production process.
3. *Management*: Concerns around all persons who have some measure of control over the process, such as employee participation, operating procedure, and management of change.

Patel (2012) advised that the PSM elements may be grouped into three main themes, *facilities, personnel,* and *technology*, as illustrated in Figure 3.2. McBride and Collinson (2011) advised of a similar packaging of PSM elements into categories of *people, process,* and *plant*.

Lutchman et al. (2012) categorized the PSM elements into *people, processes and systems*, and *facilities and technology*, as shown in Table 3.1. PSM therefore

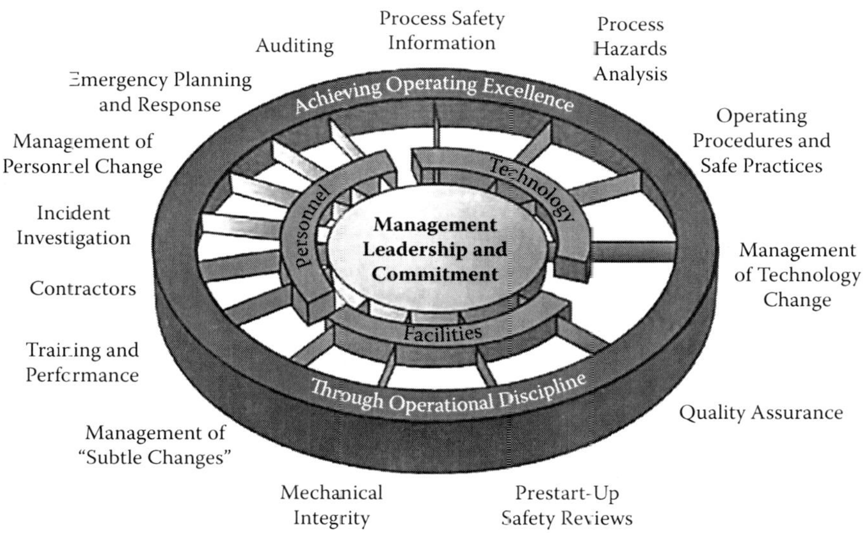

FIGURE 3.2
PSM elements categorized into facilities, personnel, and technology. (From Patel, J., The Effects of Risk Management Principles Integrated with Process Safety Management (PSM) Systems, presented at 62nd Canadian Society Chemical Engineering Conference, Incorporating ISGA-3, Vancouver, BC, Canada, October 14–17, 2012.)

TABLE 3.1

OSHA PSM Requirements—People, Processes, and Systems, and Facilities and Technology Requirements

Management Leadership and Organizational Commitment			
Safety Management System			
OSHA PSM Requirements	**People**	**Processes and Systems**	**Facilities and Technology**

• Employee training and competency.
• Contractor safety management.
• Incident investigation
• Management of change (People).
• Emergency preparedness planning and management.
• Management of change (Engineered/Nonengineered).
• Nonroutine work authorizations.
• Pre-startup safety review.
• Compliance audits
  • Planning.
  • Staffing.
  • Conducting the audit.
  • Evaluation and corrective actions.
• Process Safety Information
  • Hazards of the chemicals used in the processes.
  • Technology applied in the process.
  • Equipment involved in the process.
  • Employee involvement.
• Process Hazard Analysis
• Operating procedures.
• Mechanical integrity of equipment.
  • Process defenses.
  • Written procedures.
  • Inspection and testing.
  • Quality assurance.
	• Employee training and competency
	
	• Contract or Safety Management
	
	• Incident investigations
	
	• Management of change (Personnel)
	
	• Emergency preparedness, planning and management
		• Management of Engineered Change and NonEngineered Change
		
		• Nonroutine work authorization
		
		• Prestartup safety reviews
		
		• Compliance audits
		  • Planning.
		  • Staffing.
		  • Conducting the audit.
		  • Evaluation and corrective actions.
			• Process Safety Information
			  • Hazards of the chemicals used in the processes.
			  • Technology applied in the process.
			  • Equipment involved in the process.
			  • Employee involvement.
			
			• Process Hazard Analysis
			
			• Mechanical integrity of equipment.
			  • Process defenses.
			  • Written procedures.
			  • Inspection and testing.
			  • Quality assurance.
			
			• Operating Procedures

Source: Lutchman, C. et al., *Safety Management: A Comprehensive Approach to Developing a Sustainable System*, Taylor & Francis, CRC Press, Boca Raton, FL, 2012.

extends beyond a quality focus to address people, facilities, and technological requirements of an organization.

While PSM requirements are regulatory in the United States, the benefits of PSM have become abundantly clear to many countries. As a consequence, many organizations in Canada and other parts of the world have embraced the OSHA requirements of PSM and actively begun implementing PSM across their businesses. Indeed, like many of their U.S. counterparts, these organizations have gone a step further to embrace OEMS as their management system of choice—which extends beyond PSM requirements.

3.3.4 Elements of PSM

According to OSHA (2010), a PSM program must be systematic and holistic in its approach to managing process hazards and must consider the following:

1. Process design
2. Process technology

3. Process changes
4. Operational and maintenance activities and procedures
5. Nonroutine activities and procedures
6. Emergency preparedness plans and procedures
7. Training programs
8. Other elements that affect the process

The OSHA 29 CFR Part 1910-119 requirements of PSM for each element within organizations are generally supported with standards and may be summarized as follows:

1. Ensure process safety information (PSI) is available. Organizations must develop and maintain written and accessible safety information that identifies hazards associated with workplace chemicals and processes, equipment, and technology, and people required to work with these hazards.
2. Process safety hazards (process hazard analysis (PHA)) identification. Organizations should perform a workplace hazard assessment, including, as appropriate, identification of potential sources of process hazards within a process facility that can result in catastrophic consequences in the workplace. The process must also identify the extent of the hazards and risk exposures and do the following:
    a. Establish a process for addressing the hazards and risks exposures.
    b. Steward the process to ensure timely resolution of exposures.
    c. Document resolutions and actions to be undertaken.
    d. Develop a written schedule for completion.
    e. Communicate the actions to stakeholders—employees, and contractors whose work assignments may be in this process and who may be affected by the risk mitigation actions.
3. Employee involvement and participation. Consult with employees and their representatives on the development and conduct of hazard assessments and the development of chemical accident prevention plans. They must provide access to these and other records required under the standard.
4. Hot work permit. Control of work and permitting of nonroutine work is required. Organizations will adopt and apply a permitting system that allows for identification and control of workplace hazards before work is undertaken. More specifically, OSHA requires

the use of hot work permits for hot work operations conducted on or near a covered process.

5. Operating procedures. Organizations must develop and implement written operating procedures for process operations and for the chemical processes. Procedures should include operating limits and health and safety considerations.
6. Worker training and competency assurance. Organizations should provide written safety and operating information for employees and employee training in operating procedures. Emphasize hazards and safe practices that must be developed and made available.
7. Contractor safety management. Organizations must ensure contractors and contract workers are trained and competent to do any assigned work, and are provided with access to all relevant information required for performing assigned work safely.
8. Emergency preparedness planning and management. Organizations must train and educate employees and contractors in emergency response procedures. Emergency preparedness plans and drills must be maintained to ensure that during an unforeseen event, personnel know how to respond and evacuate safely.
9. Quality assurance (QA). Organizations must establish a quality assurance program to ensure that initial and replacement process-related equipment, maintenance materials, and spare parts are fabricated and installed consistent with design specifications.
10. Mechanical integrity (MI) of equipment. Organizations must establish maintenance systems for critical, process-related equipment and machinery. Written procedures are required for doing such work, and the work must be undertaken by trained, qualified, and competent workers. Inspection and testing of such equipment on an approved schedule is required to ensure the ongoing mechanical integrity of equipment and machinery and interconnecting systems.
11. Prestart-up safety reviews (PSSRs). Organizations must develop a process for conducting prestart-up safety reviews of all newly installed or modified equipment. PSSRs are to be done by qualified and competent personnel, and all deficiencies categorized as high risk are to be addressed prior to start-up.
12. Management of change (MoC). Organizations must establish and implement written procedures for managing change to process chemicals, technology, equipment, machinery, personnel, and facilities. The process should also differentiate between permanent and temporary changes and address the requirements for changing personnel in critical roles.

13. Incident investigation. Investigate all incidents that result in or are near misses that could have led to a major accident in the workplace. Findings are to be reviewed and corrective actions are to be taken. Learnings should be shared across the organization to prevent a repeat of the same or similar incident.
14. Compliance audits. Organizations shall conduct a compliance audit at least once every 3 years to verify that the procedures and practices developed under the standard are adequate and are being followed by the organization.

Bingham (2008) pointed to 20 steps to process safety as per the American Institute of Chemical Engineers (AIChE) Center for Chemical Process Safety, as shown in Table 3.2.

TABLE 3.2

Steps to Process Safety—AIChE Center for Chemical Process Safety

Commit to process safety	1. Process safety culture 2. Standards, codes, regulations, and law 3. Process safety competency 4. Workplace involvement 5. Stakeholder outreach
Understand hazards and risks	6. Process knowledge management 7. Hazards identification and risk analysis
Manage risks	8. Operating procedures 9. Safe work practices 10. Asset integrity and reliability 11. Contractor management 12. Training and performance assurance 13. Management of change 14. Operational readiness 15. Control of operations 16. Emergency management
Learn from experience	17. Incident investigations 18. Measurements and metrics 19. Auditing 20. Management review and continuous improvements

Source: Bingham, K., Process Safety Management: The Elements of PSM, PROCESSWest, 2008, retrieved March 4, 2013, from http://www.acm.ca/uploadedFiles/003_Resources/Articles_in_ProcessWest/Process%20Safety%20Management%20-%20The%20Elements%20of%20PSM.pdf.

OSHA (2010) further advised that should small organizations address the following in their safety management systems, there is a greater chance that they will be compliant with regulatory requirements.

1. Process safety information (PSI):
   a. Hazards of the chemicals used in the processes
   b. Technology applied in the process
   c. Equipment involved in the process
   d. Employee involvement
2. Process hazard analysis (PHA)
3. Operating procedures
4. Employee training and competency
5. Contractor safety management
6. Prestart-up safety review
7. Mechanical integrity of equipment:
   a. Process defenses
   b. Written procedures
   c. Inspection and testing
   d. Quality assurance
8. Nonroutine work authorizations
9. Management of change (MoC):
   a. Engineered and nonengineered changes
   b. People
10. Incident investigation
11. Emergency preparedness planning and management
12. Compliance audits:
    a. Planning
    b. Staffing
    c. Conducting the audit
    d. Evaluation and corrective actions

3.4 Operational Excellence Management System (OEMS) and Its Variations

Over the past two or three decades, some of the world's leading organizations have been operating with a management system that is designed to promote strong operating and business performance. From the authors' perspective, when the management systems of these organizations were properly analyzed, seven fundamental requirements common to them were identified:

1. Leadership commitment, motivation, and accountability
2. Establishing the required elements of the management system
3. Establishing the baseline to understand where the organization sits in relation to its desired state
4. Following a plan–do–check–act model for managing work.
5. Auditing for compliance and conformance to the requirements of each element standard
6. Closing the gaps and maintaining operational discipline in doing so
7. Continuous improvements and shared learning

These fundamentals shall be discussed in detail in subsequent chapters.

Among the common elements and their variants found in an OEMS are the following:

1. Management, leadership, and organizational commitment and accountability
2. Leadership and management review
3. Legal requirements and compliance
4. Security management and emergency preparedness
5. Qualification, orientation, and training
6. Contractor management
7. Event management and learning
8. Management of personnel change
9. Process safety information
10. Process hazard analysis
11. Physical asset system integrity and reliability:
    a. Operations and maintenance controls
    b. Quality assurance
12. Operating procedure and safe work practices
13. Establishing specific, measurable, achievable, realistic, and time-bounded (SMART) goals and targets
14. Ensuring the right organizational structure, responsibility, and authority

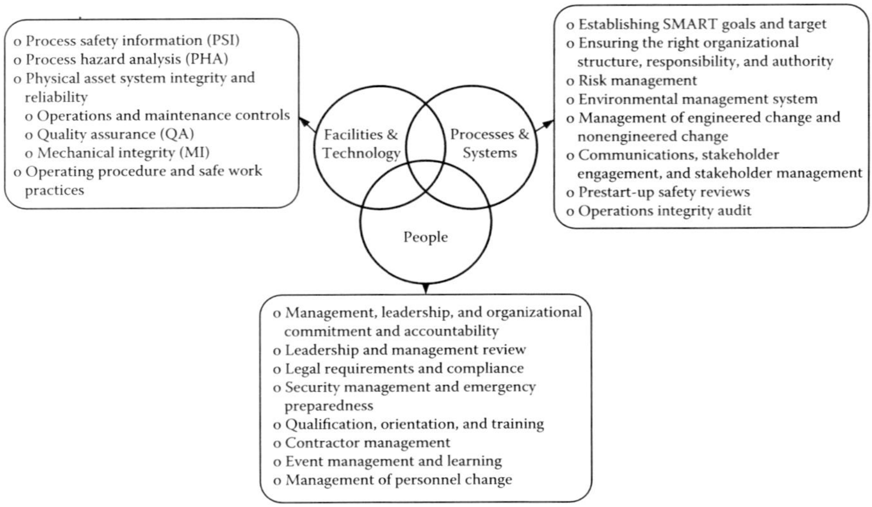

FIGURE 3.3
Categorized OEMS elements.

15. Risk management
16. Environmental management system
17. Management of engineered change and nonengineered change
18. Communications, stakeholder engagement, and stakeholder management
19. Prestart-up safety reviews
20. Operations integrity audit

Lutchman et al. (2012) categorized the requirements of a safety management system into the following:

1. People management
2. Supporting processes and systems
3. Facilities and technology

Similar categorization of OEMS elements are shown in Figure 3.3. Categorization is not a precise science and can vary from organization to organization; however, the key requirement for success in an organization is to correctly identify which elements of the OEMS are required and execute them thoroughly.

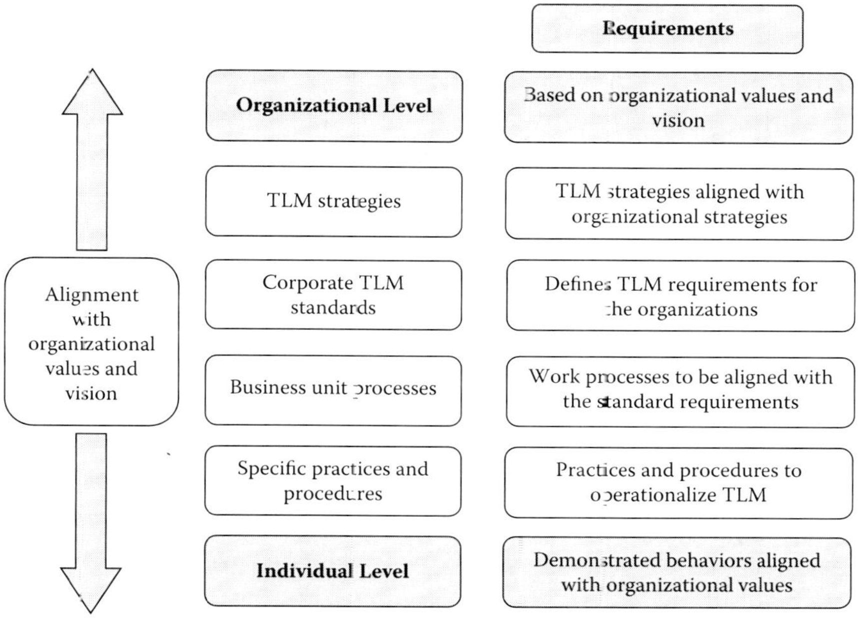

FIGURE 3.4
TLM management system and requirements. (From Suncor Energy, Total Loss Management (TLM), Suncor Sustainable Development, 2013, retrieved March 8, 2013, from http://sustainability.suncor.com/2009/pdf/rtc-tlm-outline-e-.pdf 2013.)

While there are many types of management systems, we have identified those that seemed to have surfaced as the leading management systems in today's business environment. Among the more common and less complete management systems are the following:

1. Safety management system (SMS): Focuses primarily on organizational health and safety, while other requirements of the business are managed separately from the SMS.
2. Total loss management (TLM): A management system that is very closely aligned to OEMS but may fail to recognize and include all of the required elements of an OEMS. Figure 3.4 provides an overview of TLM as applied by Petro-Canada prior to its merger with Suncor in 2009.
3. Management systems designed to meet the management requirements of a particular business activity or a specific type of organizational asset, for example, customer relationships (CRM), preventive maintenance (PMM), and materials (MM).

4. Management systems designed to meet the management requirements of all relevant areas of operation. Included among such management systems are the following:
    a. Knowledge management
    b. Financial management
    c. Human resources management
    d. Incident management

These individual management systems often work together to form a comprehensive management system when areas of overlap are removed, often leading to an OEMS.

Table 3.3 shows the evolution of management systems toward an OEMS. It also provides the evolving element required to achieve the conformance/compliance to the management system.

3.4.1 Quality Warning Signs in OEMS Organizations

Over the years, we have witnessed many companies sacrifice long-term Quality, Health, Safety, and Environment (QHSE) systems for short-term financial gains. This can be noted in the lives lost in the Massey Energy mine disaster in West Virginia, or environmental disasters such as the *Deepwater Horizon* explosion in the Gulf of Mexico, or as recent as the 2014 largest recall from Toyota in safety defects. These issues beg the question: Are there warning signs that are indicators of a change in corporate attitudes toward QHSE issues? Weisbrod (2010) has summarized five warning signs that indicate when QHSE is at risk in organizations:

1. Repeated cost-cutting cycles
2. Operational signals ignored or delayed
3. Aging equipment or degradation of maintenance services
4. Direct cuts to quality or operational excellence personnel
5. Elimination or outsourcing of customer assistance resources

Corporations can reverse this cycle by returning to the basic principles that initially led to greatly improved quality and by adhering to the fundamentals discussed in this chapter. According to Weisbrod (2010), organizations should follow the guidance provided by W. Edwards Deming and strive for significant improvements in all aspects of business activity.

TABLE 3.3
Principles and Elements of Various Management Systems

TQM—1950s	ISO 9000 and Family—1980s	Process Safety Management (PSM)—1990s	OEMS—2000s
1. Customer focused 2. Total employee involvement 3. Process centered 4. Integrated system 5. Strategic and systematic approach 6. Continual improvement 7. Fact-based decision making 8. Communications	1. Customer focused 2. Leadership 3. Involvement of people 4. Process approach 5. System approach to management 6. Continual improvement 7. Factual approach to decision making 8. Mutually beneficial supplier relationships	1. Process safety information (PSI) 2. Process hazard analysis (PHA) 3. Operating procedures 4. Employee training and competency 5. Contractor safety management 6. Prestart-up safety review 7. Mechanical integrity of equipment 8. Nonroutine work authorizations 9. Management of change (MoC) 10. Incident investigation 11. Emergency preparedness planning and management 12. Compliance audits 13. Quality assurance 14. Nonroutine work	1. Management, leadership, and organizational commitment and accountability 2. Leadership and management review 3. Legal requirements and compliance 4. Security management and emergency preparedness 5. Qualification, orientation, and training 6. Contractor management 7. Event management and learning 8. Management of personnel change 9. Process safety information (PSI) 10. Process hazard analysis (PHA) 11. Physical asset system integrity and reliability • Operations and maintenance controls • Quality assurance (QA) • Mechanical integrity (MI) 12. Operating procedure and safe work practices 13. Establishing SMART goals and targets 14. Ensuring the right organizational structure, responsibility, and authority 15. Risk management 16. Environmental management system 17. Management of engineered change and nonengineered change 18. Communications, stakeholder engagement, and stakeholder management 19. Prestart-up safety reviews 20. Operations integrity audit

References

Bingham, K. (2008). Process safety management: The elements of PSM. PROCESSWest. Retrieved March 4, 2013, from http://www.acm.ca/uploadedFiles/003_Resources/Articles_in_ProcessWest/Process%20Safety%20Management%20-%20The%20Elements%20of%20PSM.pdf.

Dale, B.G., van der Wiele, T., and van Iwaarden, J. (2007). *Managing quality*. 5th ed. Blackwell Publishing Ltd., Oxford.

Folaron, J. (2003). The Evolution of Six Sigma. *Six Sigma Forum Magazine*, 2(4), 38–44. Retrieved April 7, 2014, from http://asq.org/pub/sixsigma/past/vol2_issue4/folaron.html.

Hashmi, K. (2010). Introduction and implementation of total quality management (TQM). Retrieved February 20, 2013, from http://www.isixsigma.com/methodology/total-quality-management-tqm/introduction-and-implementation-total-quality-management-tqm/.

Honarpour, A., Jusoh, A., and Khalil, M.N. (2012). Knowledge management, total quality management and innovation: A new look. *Journal of Technology Management and Innovation*, 7(3), 22–31. Retrieved February 21, 2013, from EBSCOhost database.

Khasawneh, S.N., Al-Hashem, A.O., and Al-Zoubi, W.K.A. (2012). Application of total quality management systems (TQMS) and its impact on competition policy in industrial plants: An empirical study on facilities on industrial sector in Jordan. *Far East Journal of Psychology and Business*, 8(1), 1–22. Retrieved February 21, 2013, from EBSCOhost database.

Lutchman, C. (2010). *Project execution: A practical approach to industrial and commercial project management*. Taylor & Francis, CRC Press, Boca Raton, FL.

Lutchman, C., Evans, D., Maharaj, R., and Sharma, R. (2013). *Leveraging networks and communities of practice for continuous improvement*. Taylor & Francis, CRC Press, Boca Raton, FL.

Lutchman, C., Maharaj, R., and Ghanem, W. (2012). *Safety management: A comprehensive approach to developing a sustainable system*. Taylor & Francis, CRC Press, Boca Raton, FL.

Majstorović, V.D., and Marinković, V.D. (2012). Research of the impact of quality management principles on integrated management systems practice in Serbia. *Acta Technica Corvininesis—Bulletin of Engineering*, 5(1), 143–149. Retrieved February 21, 2013, from EBSCOhost database.

McBride, M., and Collinson, G. (2011). Governance of process safety within a global energy company. *Loss Prevention Bulletin*, 217, 15–25. Retrieved November 3, 2012, from EBSCOhost database.

Occupational Safety and Health Administration (OSHA). (2000). Process safety management. OSHA 3132. U.S. Department of Labor Occupational Safety and Health Administration. Retrieved August 23, 2012, from http://www.osha.gov/Publications/osha3132.pdf.

Occupational Safety and Health Administration (OSHA). (2010). Process safety management guidelines for compliance. U.S. Department of Labor Occupational Safety and Health Administration. Retrieved January 1, 2013, from http://www.osha.gov/Publications/osha3133.html.

Patel, J. (2012). The effects of risk management principles integrated with Process Safety Management (PSM) Systems, presented at 62nd Canadian Society Chemical Engineering Conference, Incorporating ISGA-3, Vancouver, BC, Canada, October 14–17.

Schultz, J.R. (2013). Out in front. *Quality Progress*, 19–23.

Shirshendu, R., and Prasun, D. (2011). Quality improvement initiatives for support functions in an industry: Two cases. *International Journal for Quality Research*, 5(3), Retrieved June 21, 2014, from https://www.researchgate.net.

Suncor Energy. (2013). Total loss management (TLM), Suncor sustainable development. Retrieved March 8, 2013, from http://sustainability.suncor.com/2009/pdf/rtc-tlm-outline-e-.pdf.

Weisbrod, S. (2010). Warning! Warning! Five signs that quality is at risk in your organization. *Quality Progress*, 10–11.

4

Benefits of a Management System

There are many benefits from organizations adopting and maintaining an effective management system. Organizations that apply and work with management systems often generate excellent business performance for their stakeholders. Typically, such organizations demonstrate

- Standardized business processes and practices, often guided by requirements established in elements, standards, policies, and procedures
- Operational discipline that is reflected in consistent approaches to business practices and their way of doing business
- Strong business reliability and operations performance that lead to strong production and financial performance
- Work that is often conducted in an organized and planned manner
- Excellent environmental and social responsibility performance
- Motivated workforce that is reflected in less employee turnover
- Employer of choice—high employee morale
    - Lower worker turnover—free to go, want to be there
    - Attracts the best and brightest
    - High worker morale and excellent work ethics and attitudes
- High productivity and profitability—standardization and elimination of duplicated efforts
- Fewer incidents
- High operating reliability
- Greater operating efficiency
- More efficient use of scarce resources
- Maximized value creation
- Goodwill for environment and socially responsible behaviors
    - Minimizes impact to the environment
    - Returns to society in the forms of employment, community development, and disaster support
    - Excellent corporate goodwill and image

- More effective knowledge management/transfer and consistency in decision making
    - Creates a learning culture
    - Fewer repeat incidents
    - Better and more timely value-added decisions
    - Quality knowledge shared
    - Applied knowledge
- Improved customer, contractor, and supplier relationships
    - Greater efficiency and performance from shared values
    - Contractors and suppliers know what is expected of them when working at the company
    - Less rework required
    - On-budget and on-schedule delivery of projects
- Peer and industry recognition
    - Know-how and competence
    - Best practices

Undoubtedly, therefore, management systems generate operation discipline and excellence. When the fundamentals of the organizational management system are clearly understood by all levels of the organization, and workers understand the positive impacts on their work from operating within the confines and requirements of the management system, organizational health and safety, production, and financial performance are optimized.

4.1 Case Studies: Management Systems Applied

There are many instances where the application of a management system has led to an exceptional business performance even during extenuating circumstances. The author seeks to highlight the performance of a few selected organizations that have adopted a management system approach for running their business. While the complexity of the management system adopted by each of these organizations may vary by the numbers of elements and subelements, the underlying principles remain the same, with principal focus on people, processes and systems, and facilities and technology, as categorized in Figure 3.3.

4.1.1 ExxonMobil: Operations Integrity Management System (OIMS)

Among the early pioneers of a structured and organized management system in business, ExxonMobil stewards 11 elements in its management system. Today OIMS is deployed across all of ExxonMobil's global operations for standardization and consistency in its operations. ExxonMobil (2012) advised that OIMS "forms the cornerstone of our commitment to operational excellence and provides a solid framework to achieve safe and reliable operations" (p. 22). According to ExxonMobil (2012), OIMS provides a "framework for managing the safety, security, health, and environmental risks inherent in our business, and provides the structure to ensure that we meet or exceed local regulations" (p. 22). ExxonMobil (2012) also advised of the requirement for continuous assessments of the framework and its effectiveness, from which learnings are derived for improvements. Furthermore, the organization pointed to regular assessments of its business for compliance and conformance to the requirements of OIMS as a means for performance management.

From ExxonMobil's perspective, OIMS is critical for sustaining an environment in which "operational excellence is critical to maximizing long-term shareholder value" (p. 22). ExxonMobil credits OIMS for its solid business performance and for enabling "continuous improvement in our safety performance, increased reliability, and lower operating cost" (p. 22) as reflected in Figure 4.1.

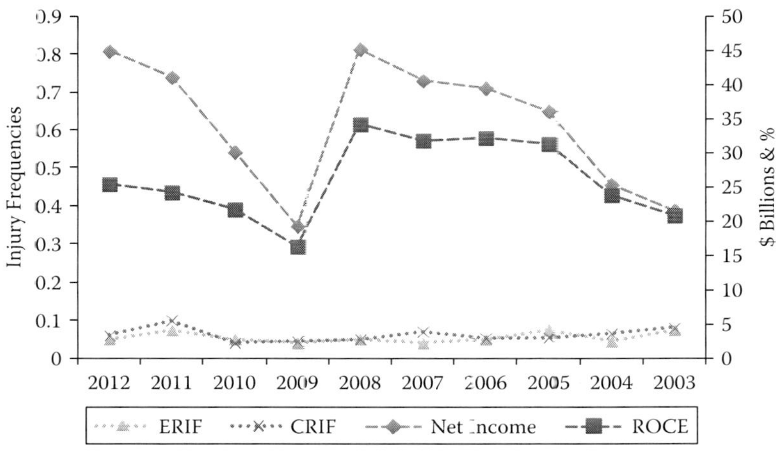

FIGURE 4.1
Business performance of ExxonMobil. (Generated from annual reports from ExxonMobil.)

4.1.2 Suncor Energy: Operational Excellence Management System (OEMS)

Suncor Energy is a Canadian integrated oil and gas producer and is among the largest organizations in Canada based on assets. The organization is headquartered in Calgary, Alberta, and is a dominant leader in both mining and steam-assisted gravity drainage (SAGD) extraction of heavy oils from the Alberta oil sands. According to Suncor Energy (2012), sometime after 2009, Suncor embarked on the activation of a management system that is aligned with international standards.

OEMS is Suncor's management system that is the framework to achieve operational excellence. It is aligned with internationally recognized management systems and is a best practice in defining and continually improving the reliability and performance of assets. By following consistent standards and processes, Suncor expects to increase positive business results by reducing risk, enhancing safety and reliability, and improving environmental performance. OEMS in Suncor is intended to

- "Ensure compliance with all legal requirements and commitments" (p. 9)
- "Adequately manage and control risk" (p. 9)
- "Apply a continuous improvement mindset to operational performance and sustain it" (p. 9)

To achieve these excellence goals, the organization depends on "integrated and consistent standards and procedures for use across the entire organization" (p. 9). In 2012, the organization had benchmarked its current position relative to the requirements of its vision (established its baseline) and was in the process of developing strategic goals to close gaps identified between current and desired states.

To Suncor, operational excellence means operating in a way that is safe, reliable, cost-efficient, and environmentally responsible. The core focus of OEMS, according to Suncor, is as follows:

People: "Recruit and retain sufficient, capable, motivated people and significantly improve the people productivity of our business" (Suncor Energy, 2012, p. 9).

Personal and process safety: Continue our journey to zero and significantly improve the integrity and performance of our assets.

Reliability: Significantly improve the reliability of our business.

Environmental excellence and sustainability: "Significantly improve the environmental performance of our business and go 'beyond compliance' in key areas" (p. 9).

TABLE 4.1

OEMS Provides Controls to Eliminate the Causes of Unplanned Events and Incidents

Performance measurement	Four sources of risk	A finite number of causes for unplanned events and incidents	Controls to address causes and reduce incidents	Suncor controls OEMS
People Personal & Process Safety Sustainability Reliability	People Processes Equipment Change	Examples Expectations don't exist Lack of knowledge Wrong incentives Equipment not capable Personnel not allocated Process not capable Management of change inadequate	Examples Procedures Training/certification Performance management Engineering disciplines Planning/scheduling Management of change process Organization structure Contractor prequalification Root cause analysis Site response Document control Hazard assessment	18 OEMS Elements

Source: Adapted from Suncor Energy, Operational Excellence Management System, 2014, retrieved September 19, 2014, from http://sustainability.suncor.com/2013/en/about/operational-excellence-management-system.aspx.

These four operational excellence strategic goals provide focus and direction for Suncor and help mitigate against risks (Suncor Energy, 2014). Table 4.1 illustrates how OEMS provides controls to eliminate the cause of unplanned events and incidents. Table 4.2 illustrates Suncor's 18 OEMS elements and the summarized objectives within the organization of each element. Suncor categorizes its 18 elements into the plan–do–check–act (PDCA) framework that allows the organization to better manage the OEMS process within Suncor.

TABLE 4.2

Suncor's OEMS Elements and Element Objectives

	Elements	Objectives
Plan	1. Leadership and accountability	Establishes expectations for leadership in implementing and maintaining OEMS, and clear accountabilities for the performance and continuous improvement of OEMS.
	2. Risk management	Establishes expectations for implementing a systematic approach to identifying and managing risk through the use of standardized tools and processes.
	3. Legal requirements and commitments	Requires that applicable legal requirements and commitments are identified, interpreted, and translated for action and relevance.
	4. Goals, targets, and planning	Sets requirements for setting goals and targets to develop business plans and to assist in the understanding of expected contributions, priorities, and deliverables.
	5. Management of change	Outlines requirements for managing organizational and operational change to minimize risk introduced by these changes. Includes changes managed internally and by third parties.
Do	6. Structure, responsibility, and authority	Establishes requirements for the development of the structure, responsibilities, and authorities to meet OEMS requirements and support the operational excellence strategy
	7. Learning and competence	Provides the framework to identify competency requirements, assess competency, identify and implement learning activities, and maintain the collective competence in work areas.
	8. Asset life cycle	Outlines the processes and systems necessary for effective asset life cycle management to ensure safe and reliable asset development and management.
	9. Operations and maintenance controls	Outlines a systematic approach for identifying the required controls and methods, and for monitoring adherence to these controls, in order to ensure the risk associated with the maintenance of operations is managed effectively.
	10. Contractor management	Establishes requirements for implementing a systematic contractor management program ensuring contractors are evaluated and selected, and that they perform in a safe, environmentally sound, and cost-effective manner.

TABLE 4.2 (CONTINUED)

Suncor's OEMS Elements and Element Objectives

	Elements	Objectives
	11. Data, document, and information management	Defines the requirements for the identification, control, and management of all data, documents, and information deemed to be critical for Suncor operations.
	12. Emergency management	Establishes requirements for the effective planning and response related to emergencies, including crisis management, operational and security emergency management, and business continuity.
	13. Communication and stakeholder relations	Outlines the framework for a systematic approach to the management of communications and stakeholder relations with employees and external stakeholders.
Check	14. Quality assurance	Sets out expectations for the implementation of quality assurance systems to ensure business processes, products, and services meet internal and external quality assurance expectations.
	15. Incident management	Outlines the formal processes required for the reporting, investigation, and subsequent management of incidents and hazards.
	16. Audit and assessments	Establishes clear requirements for the implementation and maintenance of audit and assessment processes at various levels of the organization.
	17. Corrective actions	Outlines the requirements to establish processes for development, management, and stewardship of corrective actions.
Act	18. Management review	Sets the requirements for putting processes in place for senior management to conduct documented reviews of OEMS to ensure continuing suitability, adequacy, and effectiveness.

Source: Adapted from Suncor Energy, Operational Excellence Management System, 2014, retrieved September 19, 2014, from http://sustainability.suncor.com/2013/en/about/operational-excellence-management-system.aspx.

Figure 4.2 provides an overview of Suncor's business performance. Employee recordable injury frequency (ERIF) and Contractor Recordable Injury Frequency (CRIF) data prior to 2007 were not publicly available for Suncor.

Noteworthy, too, both Suncor and Petro-Canada announced and executed a merger in 2009–2010. While 2010 onward reflected data of the merged entity, the authors were unable to validate that prior data reflected that of the merged entity.

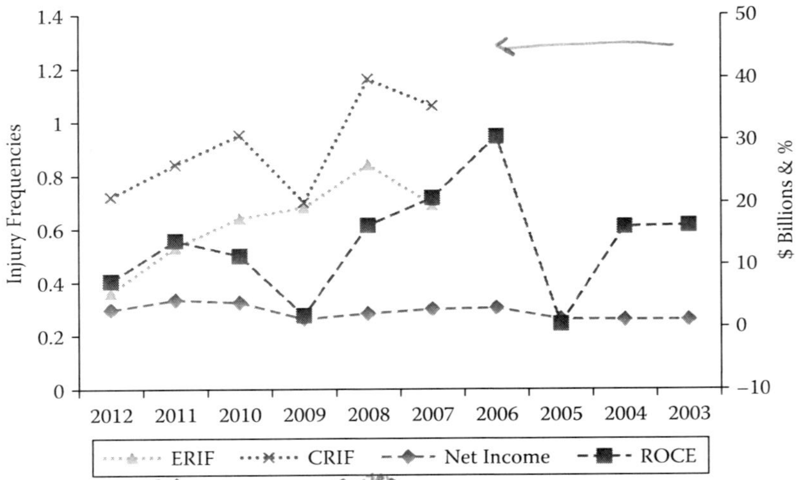

FIGURE 4.2
Business performance of Suncor Energy. (Generated from Suncor's annual reports and website tools, http://www.suncor.com.)

4.1.3 Beyond Petroleum (BP): Operations Management System (OMS)

Introduced in 2008, BP (2008) advised that its OMS has two main purposes: "to further reduce HSSE risks in our operating activities and to continuously improve the quality of those operating activities" (p. 2). According to BP (2008), OMS provides a framework of processes, standards, and practices to help deliver consistent performance, progressing to excellence in operations and safety. Sieg (2007) suggested that an OMS helps the organization establish priorities, assess the organization's performance, engage and involve its people, and maintain the integrity of its operations.

OMS in BP comprises 8 elements and 47 subelements (Sieg, 2007, p. 18) that are categorized into people, performance, process, and plant. The goal of a fully implemented OMS program in BP is to deliver safe, responsible, and reliable operations that can be continuously improved upon. OMS in BP guides the local OMS (LOMS) of each global operating entity of the organization, "providing an integrated, consistent way of operating" (p. 7). Sieg (2013), Group Head of Operations, Safety and Operational Risk, BP, advised:

> Our OMS is designed to drive a rigorous and holistic approach to safety, risk management and operational integrity. It provides considerable detail describing what we expect, and what good performance looks like, yet it is built around a handful of simple operating principles and concepts. Most importantly, OMS is designed to help leaders focus on the few things that are most important when delivering safe, compliant and reliable operations. I've seen great things happen when leaders use

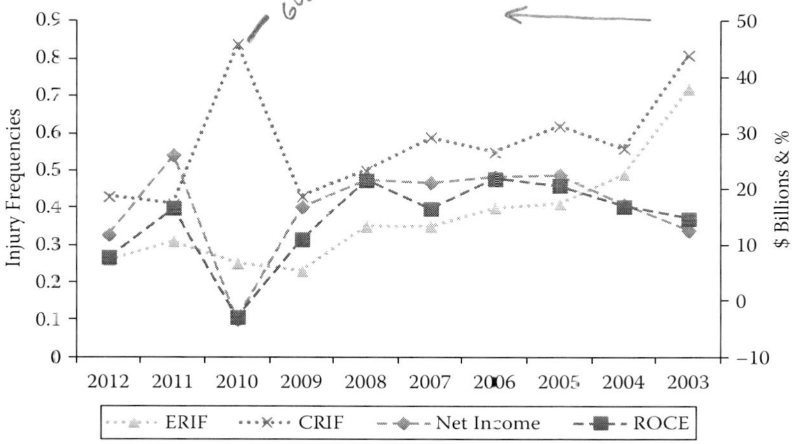

FIGURE 4.3
Business performance of Beyond Petroleum. (Generated from BP's annual reports and safety data and mapping tools, http://www.bp.com/sectionbodycopy.do?categoryId=9048973&contentId=7083700.)

> the simplicity of OMS to clarify and establish their operating priorities and expectations. The principles of OMS are fundamental to how we deliver safe operations at BP. (p. 7)

Figure 4.3 provides an overview of BP's business performance. As noted in this figure, 2010 shows the impact of the Gulf of Mexico spill (Macondo incident), which resulted in 11 fatalities and billions of dollars of environmental damage and clean up cost.

4.1.4 Chevron: Operational Excellence Management System (OEMS)

Operational excellence (OE) is a critical driver for business success and a key component of Chevron's enterprise execution strategy. According to John Watson, current chairman and CEO of Chevron, the OEMS is a comprehensive, proven means for the systematic management of process safety, personal safety and health, the environment, reliability, and efficiency, which provides Chevron with a competitive advantage and drives business results (Chevron, 2014). Chevron's operational excellence management system consists of three main parts:

- *Leadership accountability*: According to Chevron (2014), success in OEMS depends largely on leadership. Leaders are required to create a shared vision and prioritize goals and stewardship of progress on plans that focus on the highest impact items. Leaders are required to demonstrate genuine care for the welfare of all workers and to protect the environment with a risk mitigation mindset for process safety risks.

- *Management system process*: Chevron's management system process (MSP) is a "systematic approach used to drive progress toward world-class performance" (Chevron, 2014). Linked to the business planning process, the MSP begins with a clear vision of what success looks like and assesses the current state with the desired vision state. Gaps identified between current and desired states are prioritized, planned, developed, and executed for gap closure.
- *Operational excellence (OE) expectations*: Chevron (2014) identified 13 elements in its management system for achieving operational excellence. These elements define and detail corporate expectations that are achieved through supporting processes and standards established by the local operating entity.

OE expectations are organized under 13 elements (Table 4.3) and detail specific requirements for the management of safety, health, environment, reliability, and efficiency. The OE expectations are met through processes and standards put in place by local management. Standards specify requirements to satisfy OE expectations, and processes also specify a systematic approach regarding how to manage the requirements (Chevron, 2014). According to Chevron (2014), by applying the highest priority to the health and safety of its workforce, combined with assets and environment protection, OEMS translates this priority into operational excellence, competitive advantage, and strong business performance. Figure 4.4 provides an overview of Chevron's health and safety performance as well as its business performance.

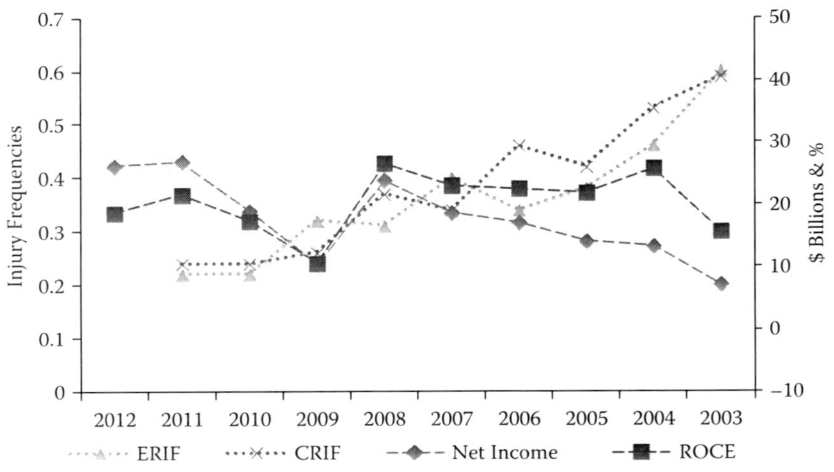

FIGURE 4.4
Business performance of Chevron. (Generated from annual reports and sustainability reports, 2003–2011.)

TABLE 4.3

Elements of Various Reviewed Excellence Management Systems

ExxonMobil	Suncor Energy	Beyond Petroleum (BP)	Chevron
1. Management, leadership, commitment, and accountability	1. Leadership and accountability	1. Leadership	1. Security of personnel and assets
2. Risk assessment and management	2. Risk management	2. Results	2. Facilities design and construction
3. Facilities design and construction	3. Legal requirement and commitments	3. Privilege to operate	3. Safe operations
4. Information/documentation	4. Goals, targets, and planning	4. Risk	4. Management of change
5. Personnel and training	5. Management of change	5. Procedures	5. Reliability and efficiency
6. Operations and maintenance	6. Structure, responsibility, and authority	6. Assets	6. Third-party services
7. Management of change	7. Learning and competence	7. Optimization	7. Environmental stewardship
8. Third-party services	8. Asset life cycle	8. Organization	8. Product stewardship
9. Incident investigation and analysis	9. Operations and maintenance controls	47 supporting subelements	9. Incident investigation
10. Community awareness	10. Contractor management		10. Community and stakeholder engagement
11. Emergency preparedness	11. Data, document, and information management		11. Emergency management
	12. Emergency management		12. Compliance assurance
	13. Communication and stakeholder relations		13. Legislative and regulatory advocacy
	14. Quality assurance		
	15. Incident management		
	16. Audits and assessments		
	17. Corrective actions		
	18. Management review		

Table 4.3 compares elements of the management systems of the four organizations examined in this chapter. There are many commonalities among these management systems. However, when the detailed subelements are examined, they all appear to fall into the categories identified in Figure 3.3 in Chapter 3: people, processes and systems, and facilities and technology.

In subsequent chapters the authors detail subelements of each element of an OEMS to provide readers with a better understanding of how each element/subelement works in conjunction with others to generate a complete management system that is designed to provide an exceptional model for achieving safe, reliable, and exceptional operations and business performances.

4.2 Conclusions

These elements and subelements, when appropriately identified and properly rolled out and supported across organizations, will provide the framework required for standardized and consistent operating principles complemented with progressive worker behaviors and business culture. Consistent with the values derived from a world leading business performer, ExxonMobil, OEMS will provide a "framework for managing the safety, security, health, and environmental risks inherent in our business, and provides the structure to ensure that we meet or exceed local regulations" (ExxonMobil, 2012, p. 22).

Similarly, according to Chevron, "success in operational excellence requires discipline in both the planning and execution of work necessary to manage safety, health, environment, reliability and efficiency with world-class results" (Chevron, 2014). Steve Williams, president and CEO of Suncor, in the 2013 annual report (Suncor Energy, 2013, p. 3), stated that "safe, reliable and environmentally responsible operations are integral to our success. Operational excellence is about doing the right work, the right way, every time—and I believe it's key to creating sustainable long-term shareholder value."

References

Beyond Petroleum (BP). (2008). The BP operating management system framework. Part 1. An overview of OMS. GFD 0.0-0001. Version 2. November 3. Retrieved May 19, 2013, from http://ecbaku.com/file/hse/OMS_Framework.pdf.

Chevron. (2014). Operational excellence management system: An overview of the OEMS. Retrieved May 5, 2014, from http://www.chevron.com/about/operationalexcellence/managementsystem/.

ExxonMobil. (2012). 2012 financial and operating review. Retrieved April 13, 2013, from http://thomson.mobular.net/thomson/7/1465/4737/document_0/F&O_as%20published_032013_print.pdf.

Sieg, J. (2007). Driving operating excellence across an organization. Retrieved May 19, 2013, from http://www.strath.ac.uk/Other/cpact/presentations/2007/pdfs/sieg.pdf.

Sieg, J. (2013). BP and sustainability: How we operate. Retrieved May 19, 2013, from http://www.bp.com/sectiongenericarticle800.do?categoryId=9048945&contentId=7082799.

Suncor Energy. (2012). 2012 annual report. Retrieved April 16, 2013, from http://www.suncor.com/pdf/Suncor_Annual_Report_2012_en.pdf.

Suncor Energy. (2013). 2013 annual report. Retrieved May 6, 2013, from http://www.suncor.com/pdf/Suncor_Annual_Report_2013_en.pdf.

Suncor Energy. (2014). Operational excellence management system. Retrieved September 19, 2014, from http://sustainability.suncor.com/2013/en/about/operational-excellence-management-system.aspx.

Section II

Fundamentals of an Operational Excellence Management System

5

Introduction: Fundamentals of an Operational Excellence Management System

In this section, the authors identify and discuss seven fundamentals essential for organizations to achieve operational excellence in their business. These fundamentals, although not new, limit the potential for the organization to achieve excellence if not properly and adequately addressed. To achieve operational excellence, businesses must take a disciplined approach toward ensuring that these fundamentals are addressed in a sustained and consistent manner across the business. Failure to adopt a disciplined approach to addressing these fundamentals can result in substandard health, safety, and environmental performance, as well as weaker business performance.

It is clear that workplace incidents and accidents can significantly impact the financial performance and goodwill of an organization. Indeed, one needs only to look at the history of Beyond Petroleum (BP) to appreciate the impact of failing to follow a disciplined approach to its management system. Incidents such as the 2005 Texas City BP explosion and the 2010 Gulf of Mexico oil spill, with their accompanying business and financial impacts, are strong indicators of a lack of operational discipline, and they highlight the consequences for such indiscipline.

Operational excellence does not occur overnight, nor can it be purchased off the shelf and applied across the business for a sustainable outcome. It is the outcome of a strategic decision adopted by an organization to execute and sustain a management system. It can be quite costly and must be supported over an extended period for it to take root and be accepted in the organization. Once the principles of the management system are accepted by all workers of the organization, an operationally excellent, self-sustaining management system evolves over time. *This self-sustaining management system eventually becomes a part of the culture of the organization that is aligned with its values and how we do business.*

The authors list the seven sequential and essential fundamentals for an operational excellence management system as follows:

1. Leadership commitment and motivation
2. Identifying and executing applicable and relevant elements to achieve operations excellence in your business
3. Establishing the baseline for your business

4. Closing the gaps identified through demonstrated disciplined behaviors
5. Following a simple business management model, namely, plan-do-check-act
6. Auditing for compliance to both regulatory and legal requirements and conformance to internal policies, procedures, and practices
7. Seeking to improve continuously and to capture and share learning on an ongoing basis

Each of these fundamentals is discussed in detail to demonstrate how it impacts the organization's ability to achieve operational discipline and excellence.

Well supported with graphics and practical examples, the authors provide a simple roadmap for organizations to evolve their management system into an operational excellence management system, designed to reduce workplace incidents and improve business performance on a sustainable basis. The management system principles discussed in this section are intended for the business leader who is motivated to transition his or her organization from ordinary, through best in class, to an organization of world-class stature and performance.

6

Fundamental 1: Leadership Commitment and Motivation

The authors identify leadership commitment and motivation as the first and, perhaps, most important fundamental essential for creating and sustaining an operational excellence management system (OEMS). In the absence of leadership commitment and motivation, operational excellence will continue to be an elusive goal that is pointless in pursuing. Leadership at all levels of the organization must be committed and motivated for success in driving toward operational discipline and excellence.

Implementing and sustaining an OEMS takes place over an extended period that may often span 5–10 years before the management system takes firm hold across the organization. There can be no faltering during this period since perceived or actual indecisions relating to continued support of the management system can result in tremendous setbacks to the implementation and sustainment processes. As a consequence, therefore, strong, committed, and motivated leadership is required for the successful implementation and sustainment of an OEMS.

Visionary, transformational leadership skills and behaviors are essential to transition the management system of an organization from ordinary to disciplined and operational excellence. In view of this, therefore, leaders should possess well-developed transformational leadership skills in the following areas:

- Creating a shared vision
- Possessing strong capabilities in promoting involvement, consultation, and participation
- Encouraging creativity, innovation, proactivity, responsibility, and excellence
- Having moral authority derived from trustworthiness, competence, sense of fairness, sincerity of purpose, and personality
- Leading through periods of change, challenges, ambiguity, and intense competition or high-growth periods
- Promoting intellectual stimulation and considering capabilities of each worker

- Displaying a willingness to take risks and generate and manage change
- Being culturally sensitive and leading across cultures and international borders
- Building strong teams while focusing on macro-management
- Exemplifying charisma and motivating workers to strong performance by inspiring their hearts and minds

6.1 Creating a Shared Vision

A vision is essentially an aspiration of what the organization should seek to achieve in the long term. A vision acts as a beacon toward which the organization is headed and provides clear guidance on desired actions to all workers of the organization. The vision of the organization is supported by its mission statement, which determines what the organization intends to achieve in the medium term. The mission of the organization must be clearly understood by all workers of the organization and supported by the strategic goals and resourcing of the organization. The strategic goals of the organization are supported by its short-term annual plans and tactical plans, which are flexible and accommodating to business environment changes and fluctuations. A common thread among the organizational vision, mission, strategic goals, and annual and tactical plans is the need for full alignment.

Figure 6.1 provides an overview of plans, focus, and actions supporting the mission and vision of the organization. As demonstrated in this figure, the strategic 10- to 15-year plan reflects the vision of the organization. Similarly, the mission of the organization is reflected in the annual plans and the 1- to 3-year business plan. These plans are generally flexible and very responsive to changes in the business environment.

A shared vision serves to uplift people's aspirations, harness and channel energy and excitement, and provide direction and motivation for all stakeholders in the organization. Creating a shared vision requires senior leadership to engage and involve people who understand the business and are capable of visioning the future of the organization. A shared vision starts with leadership at the highest level, and once the vision is crafted, there must be intense communication. Leaders must communicate, communicate, communicate, and then communicate again.

There must be strong alignment among leaders in the vision for the organization. What this means is the vision must be realistic with strong alignment with resource capabilities and growth potential of the organization. When senior leaders are not aligned with the corporate vision of the organization, it becomes very apparent in their behavior. Such leaders may voluntarily

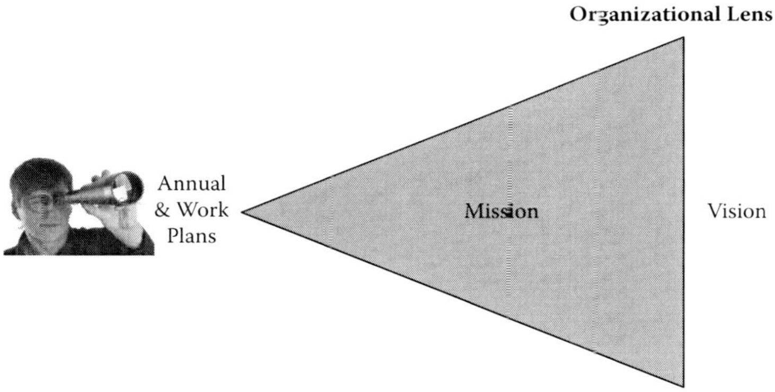

Plans	Tactical Day-to-Day Activities and Annual Plan	Strategic & Tactical 1–3 Year Business Plan	Strategic 10–15 Year Business Plan
Actions	Clear and Specific	Clear and Nonspecific	General
Focus	Individual	Business Unit/Area	Corporate

FIGURE 6.1
Differentiating among work plans, mission, and vision of the organization. (Copyright © 2013 Safety Erudite, Inc.)

leave the organization, or senior leadership may be required to take immediate action to remedy the situation either through competency assurance or removal of the leaders from the organization.

Let's try to understand the dynamics behind creating a shared vision. The following key ingredients are essential:

1. The vision must be aligned with the organizational resource capabilities and core business focus.
2. The vision must be worth aspiring for and should be achievable in the long term.
3. The vision must be carefully constructed to remove ambiguity such that all workers of the organization know where we are heading.
4. The vision should be simple to understand and explain and be easily repeated after seeing the vision statement or hearing it repeated a few times.

5. Senior leadership must be engaged and involved in creating the vision of the organization.
6. Senior leaders must sell the vision to all leaders of the organization and continue to communicate and reinforce it over an extended period. During this time, the vision is fully accepted by all workers of the organization.

When the corporate vision is shared, all workers and stakeholders are aligned, motivation is high, and business performance is strongly reflected in support of the vision.

6.2 Possessing Strong Capabilities in Promoting Involvement, Consultation, and Participation

Where OEMS is concerned, the entire workforce must be involved and made to feel a part of the process. More importantly, interpretations of the requirements for the various elements may vary and can often mean different things to different people across the various functions of the business. Consequently, leaders must be competent in promoting involvement in developing the requirements for each OEMS element, as well as in the interpretations of the requirements.

Consultation with various stakeholders and interest groups is essential to ensure their specific business needs, and requirements are met when elemental requirements are defined. Involvement, consultation, and participation lead to ownership. When workers feel a sense of ownership of the contents of each element, buy-in and support are high, and there is a much greater likelihood that requirements of the element will be met and sustained throughout the organization. In the absence of ownership, the likelihood of conformance may be reduced.

6.3 Encouraging Creativity, Innovation, Proactivity, Responsibility, and Excellence

In most instances, when executing OEMS elements across an organization, the elements will generally identify *what is required*. How to meet these requirements is generally left to those responsible for managing the process and demonstrating compliance and conformance to these requirements. In other words, *how requirements are met* is generally left to the discretion of the business unit, which will determine the value-maximizing approach for doing so.

Except in cases where the organization may have identified or adopted a proven best practice for meeting the business requirement, the desired leadership traits and behaviors are those that promote creativity, innovation, proactivity, responsibility, and excellence for meeting the elemental requirements. Leaders must therefore encourage out-of-the-box thinking and apply behaviors that are free of personal biases and stereotyping. Creative solutions, forward thinking, responsibility, and the drive for excellence are all behaviors and attitudes generated from leadership support. In the presence of other leadership behaviors and styles, these attributes in workers are stifled and curtailed.

6.4 Having Moral Authority Derived from Trustworthiness, Competence, Sense of Fairness, Sincerity of Purpose, and Personality

Moral authority derived from trustworthiness, competence, a sense of fairness, sincerity of purpose, and personality is a leadership attribute and behavior that is buried in Blanchard's ABCD model for developing and maintaining a trusting work environment. According to the Ken Blanchard Group of Companies (2010), trust in the workplace is generated from the application of the ABCD model of behaviors as described below.

Able: Demonstrate competence:

- Produce results.
- Make things happen.
- Know the organization/set people up for success.

Believable: Act with integrity—be credible:

- Be honest in dealing with people/be fair/equitable/consistent/respectful.
- Value-driven behavior "reassures employees that they can rely on their leaders" (p. 2).

Connected: Demonstrate genuine care and empathy for people:

- Understand and act on worker needs/listen/share information/be a real person.
- When leaders share a little bit about themselves, it makes them approachable.

TABLE 6.1

Leadership Behaviors That Build or Erode Trust

Erodes Trust	Builds Trust
• Lacking in communication • Being dishonest • Breaking confidentiality • Taking credit for others' work	• Giving credit • Listening • Setting clear goals • Being honest • Following through on commitments • Caring for your people

Dependable: Follow through on commitments:

- Say what you will do and do what you say you will.
- Be responsive to the needs of others.
- Being organized reassures followers.

According to the Ken Blanchard Group of Companies (2010), as leaders, you have a choice in becoming trustworthy, as shown in Table 6.1.

The Ken Blanchard Group of Companies (2010) summarized that leaders can successfully develop organizational trust in the following ways:

1. *Demonstrate trust in your people*: "If you want to create a trusting work environment, you have to begin by demonstrating trust" (p. 4).
2. *Share information*: "Information is power. One of the best ways to build a sense of trust in people is by sharing information" (p. 4).
3. *Tell it straight/never lie*: "Study after study show that the number one quality that people want in a leader is integrity" (p. 4).
4. *Create a win-win environment*: Creating competition among workers leads to a loss of trust among all.
5. *Provide feedback*: Hold regular progress meetings with direct reports. Check in and provide feedback on performance, in particular, in a timely manner to avoid surprises to workers later on.
6. *Resolve concerns head on*: Engage workers in finding solutions.
7. *Admit mistakes*: "An apology can be an effective way to correct a mistake and restore the trust needed for a good relationship" (p. 5). Some cultures have difficulty admitting mistakes. However, when mistakes are admitted, it makes the leader human and promotes greater team bonding and trust.
8. *Walk the talk*: "Be a walking example of the vision and values of the organization" (p. 5). If the leader believes in it, then so can I and so will I.

9. *Timely recognition of positive behaviors*: Recognize and reward positive behaviors in a timely manner. Such recognition must be specific and relevant. Choosing the right environment for doing so is also very important since some cultures may require pomp and show, while others may seek conservatism.

Lutchman et al. (2012) advised that "leaders should, therefore, communicate with followers in a manner that builds trust within the workforce. *Saying what you will do and doing what you said is so very important in building trust*. Workers are motivated to emulate the behaviors of leaders who make ethical and trustworthy decisions aimed at ensuring health and safety in the workplace" (p. 97).

Having moral authority suggests that followers are prepared to follow the leader because they believe in him or her. Essentially, the leader has unquestionable credibility because of consistency in behaviors, and he or she is known for doing the right thing at all times. Of course, trust in leadership and the organization is a key outcome from the demonstrated actions of the leader and employer.

Communication methods, channels, and behaviors adopted by leaders influence the levels of trust workers will place with employers (Hemdi and Nasurdin, 2006; Hopkins and Weathington, 2006). Trust is high when employers demonstrate genuine concern, empathy, and care for the health and safety of workers (Hemdi and Nasurdin, 2006). Trust in leadership is earned when leaders demonstrate competence, emotional intelligence, integrity, and ethics in decision making (Davis et al., 2000). Where the safety of personnel depends on the decisions made by leaders, such leaders must earn the trust of all workers. Care for people and cultural intelligence in a multicultural and diverse project execution environment helps in maintaining trust in leadership.

Competence in the workplace helps leaders to achieve moral authority. The ability to provide direction and support workers in achieving work plan goals provides a sense of comfort to followers that their work is supporting business needs and is valued. Leaders must therefore be competent to perform assigned leadership roles, failing which the leadership challenge in delivering an OEMS becomes larger for the entire organization.

Demonstrated fair treatment to all workers that is free of discriminatory practices and behaviors helps leaders to gain moral authority from followers. When workers feel they are treated fairly and with respect, their support for the goals of the organization is elevated. They will do more for less and can make a difference in business performance. Demonstrated unfair treatment to individuals or groups of individuals in the workplace can result in poor worker and departmental performance, and OEMS becomes more and more of an elusive dream.

When leaders demonstrate genuine commitment and sincerity of purpose for organizational deliverables, worker motivation and performance can be elevated. Lip service and weak commitment, on the other hand, are very easily detected by followers, regardless of how well a leader may try to conceal them. Once detected, followers are likely to become extremely demotivated, and apathy sets in with disastrous consequences. The message is therefore clear, and the vision must be shared, and by all levels of the organization, for leadership success in creating an OEMS.

When leaders are liked by followers, the job of influencing the behaviors of followers and guiding them is significantly easier. While personality type is not necessarily the most important factor in helping leaders gain moral authority from followers, it certainly helps. A caring, empathetic leader is more likely to gain support of followers than a goal- or task-oriented leader whose sole objective is to get the job done. Personality evaluation tools, such as Myers-Briggs and Colors, help in determining the leadership personalities that may be best suited for different roles in the execution and creation of an OEMS.

6.5 Leading through Periods of Change, Challenges, Ambiguity, and Intense Competition or High Growth

The key here is to ensure leaders are capable of managing change. Leading an organization toward an OEMS requires that leaders do the following:

- Understand and share the vision of the organization
- Clearly understand where the organization wishes to be in the future
- Analyze the gaps between the current state and desired state
- Take action to close gaps identified in a planned manner
- Prioritize and steward the proposed changes for closing gaps identified

These requirements occur while the organization continues to deliver on its business and production targets.

Change management in such a situation is quite demanding, and with multiple stakeholders across the business, this process becomes even more challenging. Ensuring the needs of various stakeholder groups are addressed (but not necessarily met) requires careful and timely engagement, involvement, consultation, and communication. In many industries, as organizations continue to compete in the quest for growth and market share, attempting to transition toward an OEMS is an increasingly difficult task. Leadership is challenged to handle multiple moving parts at the same time, and strong leadership capabilities are required to meet the challenges of this dynamic and changing work environment.

6.6 Promoting Intellectual Stimulation and Considering Capabilities of Each Worker

Leaders should seek to challenge workers by providing them with intellectually stimulating work. However, when doing so, each worker must be considered specifically to ensure that the worker is competent and capable of completing and delivering an assigned work. The key here is individual consideration. When treating people individually as opposed to lumping into a group, they become more motivated and supporting of the leader.

Making the effort to know each worker individually has tremendous payoffs for the leader. Workers feel they are relevant when the leader takes the time to meet with them individually, as well as in groups, on a frequent basis. However, such meetings require planning, take time to plan, and require commitment to follow through, as leaders juggle the many priorities in their busy lives at the office. The main outcome of such efforts is that workers feel valued and are motivated to do more for the organization when called upon.

6.7 Displaying a Willingness to Take Risks and Generate and Manage Change

As is the case in all businesses, managed risk taking is essential for success. For most organizations today, business decisions are made based on the risk tolerance of the organization as guided by a risk matrix. Leaders today must be prepared to take risks and be guided by the risk matrix of the organization. Willink (2009) suggested that leadership initiatives have been stifled within individuals. The outcome of the proliferation of heavily procedural and rules-based work environments is characterized by internal and external conformance and compliance requirements, respectively.

Willink (2009) further suggested that leaders fear "stepping out of the box, taking risk, making waves and abandoning the norm that has been set by the (corporate) status quo or common sense" (p. 120). During recessionary pressures, the fear of job losses and the accompanying difficulties in finding new employment place even more pressure on leaders to bat safely. This therefore pigeonholes them into becoming even more rule-oriented, with higher conformance to internal practices and the status quo.

The ability to generate and manage change is an absolute requirement in today's highly competitive business environment. Indeed, the challenge of continuously improving the technology, business processes, and systems requires leaders to seek opportunities for initiating continuous change. Therefore, leaders today must be able to generate and manage change in a way that creates and maximizes value for the organization.

6.8 Being Culturally Sensitive and Leading across Cultures and International Borders

Leadership today is no longer limited to regional environments characterized by homogenous work groups. The reality of a truly global workforce with immense cultural and ethnic diversity is upon us, and leaders must be able to address the work challenges associated with a truly global workforce. To become culturally sensitive, a leader must first become culturally intelligent so that he or she can recognize and address communication, collaboration, and alignment barriers that may result from cultural differences. According to Ismail et al. (2012), "cultural intelligence is the key of success in Today's world" (p. 254). They advised further that for work groups' operational effectiveness, "the group, itself, should develop cultural intelligence" (p. 254). More importantly, they added, "Dissimilar groups have greater success and greater failure potential in comparison with single-cultural groups" (p. 254).

Cultural sensitivity requires leaders to learn about the new cultures within which an organization may operate. Often, this may require the leader to live and work in the local environment within which the business operates. More importantly, leaders are required to create an inclusive work environment that is cognizant of the cultural requirements of a diverse workforce. Individual consideration and the ability to demonstrate genuine empathy and care for the entire workforce play an important role in doing so.

6.9 Building Strong Teams while Focusing on Macro-Management

How do we build strong teams? This is a challenge faced by many leaders today. Strong teams are the outcome of a careful selection of team members. The team itself should exemplify the following characteristics:

- Team members share a common goal.
- A strong team leader is selected to keep the team focused.
- Members are energized and want to be a part of the solution.
- Each member is competent and brings to the team a core capability or expertise relevant to addressing the challenge faced by the team—each member is valued.
- There is consultation, collaboration, and cooperation among members.
- Members are encouraged to challenge in a respectful way.
- Senior leadership supports are available to resource and guide the team as required.

Strong teams are required to solve complex business problems and provide solutions to challenges faced by the organization.

In a broader sense, however, where leadership is concerned, the concept of *one team* is even more important. Wherever the worker sits in the organization, regardless of whether the worker belongs to an operating asset or a functional group, his or her contributions are valued and he or she is made to feel a sense of belonging. BP (2013) claims one team as a value of the organization and defines this concept as follows: "Whatever the strength of the individual, we will accomplish more together. We put the team ahead of our personal success and commit to building its capability. We trust each other to deliver on our respective obligations within the organization" (p. 12).

Trust among members of the team is an absolute requirement. As pointed out earlier, trust is an earned attribute from consistently doing the right thing and, as discussed earlier, is rooted in the ABCD model. Building organizational and corporate teams necessarily requires a shared vision and lots of communication. Leaders must therefore commit resources required for performing these activities. With strong teams and focused areas of expertise come centers of excellence from which great opportunities for improvements may originate. Lutchman et al. (2013) discussed the implications of networks and communities of practices for generating continuous improvements in business processes and systems and overall performance.

6.10 Exemplifying Charisma and Motivating Workers to Strong Performance by Inspiring Their Hearts and Minds

There are many different motivators in the workplace for different workers. The key is leaders to know that different workers and work groups are motivated differently, and therefore individual consideration is critical for success in motivating the workforce. It is almost impossible to motivate every worker in the workforce; however, there are common actions that a leader may perform that have the same motivating impact on almost every worker. Among these are the following:

- Demonstrate genuine care and empathy for the worker.
- Show a willingness to listen (not necessarily respond and find solutions, but listen).
- Treat all workers with respect.
- Provide equal opportunities to all.
- Ensure fair treatment for all workers.
- Praise workers for work well done.

- Communicate in an unbiased, timely, and transparent way about business decisions that may impact the work in the immediate, medium, and long terms.

According to Ramlall (2004), employers should seek to respond with empathy to the personal needs and values of the employee. They should also seek to create work environments that are safe, respectful, inclusive, and productive (Ramlall, 2004; Wren, 2004).

Factors that motivate all workers at the organizational or corporate level generally include the following:

- The organization's approach to corporate social responsibilities (CSRs). More so for Generation Y today, workers want to know that the organization cares about their safety and the protection of the environment.
- How the organization addresses the personal needs of the employee.
- The work environment characteristics of the worker. Ensuring *need to have* are available is a basic minimum.
- The responsibilities and duties of the worker. Assign competent workers and develop others where gaps in their competency may exist.
- The level of supervision provided to the worker. Inexperienced and immature workers require more supervision than experienced and mature workers. Getting this wrong may lead to increased workplace incidents and accidents, and workers quitting and staying where supervision is applied but not required.
- The extent of worker effort required to perform assigned tasks. Having the right balance between automation and labor intensiveness is essential. Where work is labor-intensive, having the right numbers of workers perform the work is essential.
- The employee's perception of organizational fairness and equity. Wide gaps in the treatment between leaders and followers cause discontentment and conflict.
- Career development and advancement opportunities for all who seek them.

Employee motivation can be high when leaders focus on the health and safety of workers, are trustworthy, promote teamwork, treat workers fairly and individually, and act with empathy when communicating with workers. Leaders must say what they intend to do and do what they promised to do. Ramlall (2004) identified the following factors that leaders can address to influence worker motivation in the workplace:

1. The organizational health and safety behavior and performance. Unsafe work environments promote worker flight and reduce morale and motivation.

2. The personal needs of the employee. This may include, but is not limited to, training, personal protective equipment, or simple advice on a personal issue. Leaders must be able to recognize the body language of workers who may have unresolved personal issues and respond to them with the empathy required.
3. The work environment characteristics. Generally, commercial and industrial project environments are characterized by difficult working conditions. Unpaved roads, extreme temperatures, outdoor work, and supplemental lighting can all influence the productivity and motivation of workers.
4. The responsibilities and duties of the worker. When placed in supervisory roles, worker motivation can be affected either positively or negatively. If unprepared for the role, negative consequences may be generated. If equipped to do the job and if seen as a promotion, workers may respond positively.
5. The level of supervision provided to the worker. Immature workers will necessarily require more supervision and guidance relative to experienced workers who may require less supervision. Telling and micromanagement can lead to situations where workers *quit and stay* when they should be delegated to.
6. The extent of worker effort required to perform assigned tasks. Labor intensiveness vs. skills intensiveness influences worker motivation.
7. The employee's perception of organizational fairness and equity. Perceived unfair and biased treatment is a precursor to worker turnover and a decreased level of effort generated from workers. Workers tend to equate work with pay, and when faced with possible unfair treatment, they will find a creative means to bring about equity. Shirking on the job, resorting to absenteeism, arriving late to work, and performing poorly are methods used to level the playing field.
8. Career development and advancement opportunities. Most workers have an unspoken personal development plan for the workplace. They will be motivated to perform at high levels if career development opportunities are aligned with their personal plans.

6.11 Developing Self and Followers

Leadership in any operationally excellent environment is characteristic of worker mobility in which *all workers are free to go but want to say.* In such environments, workers are motivated, feel valued, are treated with respect, and are developed by leaders on an ongoing basis. While every worker should be developed, priority should be placed on developing supervisory and leadership personnel. There are various aspects to supervisory and leadership development in the organization that can be explored. First, employees who are entirely new to the organization must be cared for and treated accordingly. Second, a worker may be an existing employee who is new to a role and requires development in order to function efficiently.

In both instances, the four stages defined below should be considered for developing these workers. The key to success is careful attention to the duration the worker stays in each stage. As one would expect, a new worker to the organization may require a longer on-boarding process than an existing worker. Similarly, an experienced external employee may require shorter periods of mentoring and developing in the role. The development stages for the worker are as follows:

1. Stage 1: On-boarding
    - Period when the worker is new to the role
    - Ideally a prejob period exercise
    - Frontline supervisor is not yet performing supervisory work in the role
2. Stage 2: Mentoring
    - Immediate period of post on-boarding
    - Supervisor/leader is working in the role, but is under the guidance of a mentor
    - Mentor supports and assists the new supervisor/leader in the role
3. Stage 3: Developing in role
    - Frontline supervisor is growing in competence
    - Allowed to function alone on an ongoing basis
    - Continuous guidance from his or her leader is required to ensure success
4. Stage 4: Sustainment
    - Supervisor leader fully competent in the role
    - Continuous improvements
    - Consistent with the organizational values and behaviors

6.12 Situ-Transformational Leadership Model

The tenets of the situ-transformational leadership model, as discussed by Lutchman (2010), cannot be underestimated in developing the entire workforce. Lutchman (2010) superimposed transformational leadership behaviors across the various stages of Blanchard's situational leadership model to provide a simple, easy-to-follow process for developing all workers in the organization. This simplistic model proposes four stages of development similar to those identified above and applies simple workplace behaviors to move workers from immature and incompetent to mature and competent.

As the organization transforms toward an OEMS, both leaders and followers must continue to develop, learn new ways of doing things, and be able to lead and manage change. Continuous development of followers is therefore required. The situ-transformational leadership model helps leaders to achieve the development requirements of followers as per the various stages of the model.

In the quest for an OEMS, leaders should not forget that they too must continue to develop to effectively lead their workforce. Leaders must consider structured and scheduled personal development plans in order to continue to lead the workforce. The path to an OEMS is filled with change. Therefore, in this type of highly dynamic and continuously changing business environment, leaders must be continuously vigilant in self-development, or they run the risk of self-destruction in the long run. Figure 6.2 provides a model for

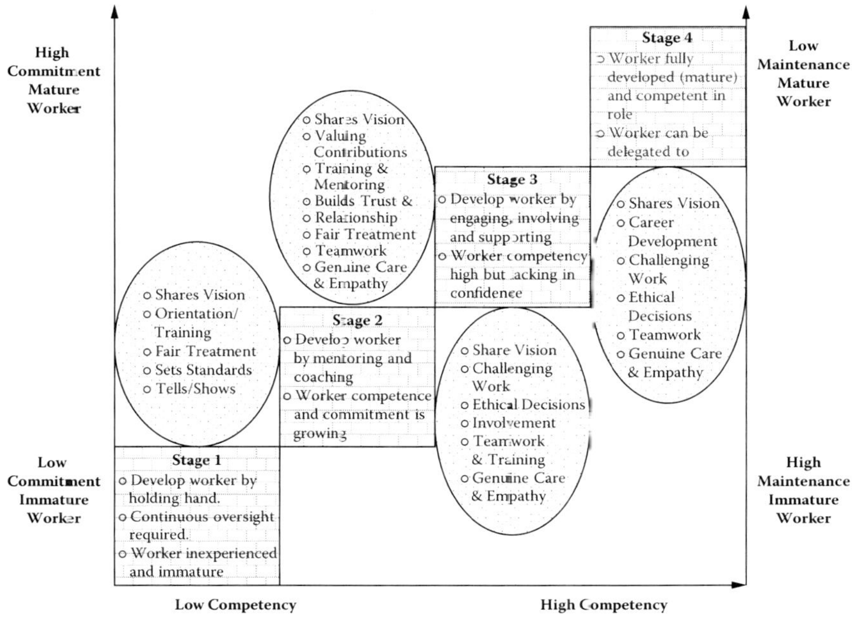

FIGURE 6.2
Situ-transformational leadership model for developing workers. (From Lutchman, C., *Project Execution: A Practical Approach to Industrial and Commercial Project Management*, Taylor & Francis, CRC Press, Boca Raton, FL, 2010.)

achieving operational excellence. The model provides a path for developing workers from immature, high maintenance, and low commitment to mature, highly motivated, and committed workers.

6.13 Leading and Managing Change When Developing an OEMS

When developing an OEMS, change is a necessary requirement. As a consequence, therefore, leaders must be able to lead and manage change. Managing change requires leaders to leverage and ensure the following requirements are met for sustainable change:

- Leadership alignment
- Stakeholder engagement
- Organizational readiness and sustainability
- Learning and capability development
- Communication
- Change network
- A culture of learning and continuous improvements

6.13.1 Leadership Alignment

When leaders are not aligned on the need for change, change is unlikely to happen or it will not occur on a sustainable basis. Figure 6.3 provides an overview of factors that contribute to nonaligned vs. aligned leadership that can impact change and change management.

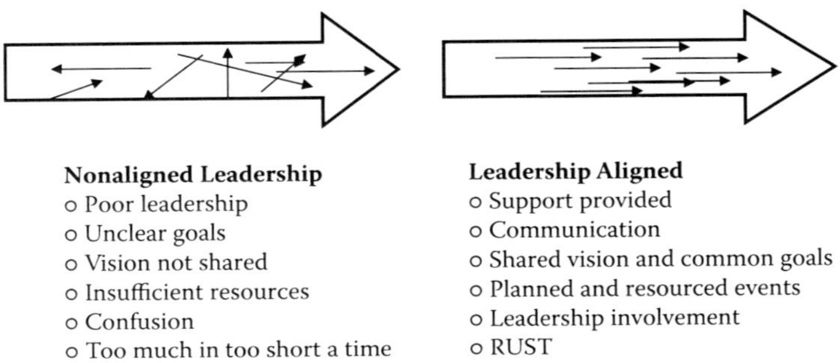

FIGURE 6.3
Factors contributing to nonaligned and aligned leadership. (Copyright ©2014 Safety Erudite, Inc.)

6.13.2 Stakeholder Engagement

When stakeholders are engaged, their concerns are heard and addressed, and sustainable change is more likely to occur. Managed change occurs when a stakeholder's impact assessment is completed and the associated risks and concerns of each stakeholder group are addressed. The key to success is engagement. Change does not require satisfaction of each stakeholder group request and requirement. However, when each stakeholder group is engaged, change is more likely to occur.

6.13.3 Organizational Readiness and Sustainability

Regardless of how important the change may be, if the organization is not ready for change, sustainable change will not occur. Organizational readiness for change requires leaders to create an environment where

- The organization is properly resourced for the proposed change.
- The change impact has been assessed and concerns are addressed and risks are mitigated to acceptable levels.
- People are engaged and feel their concerns have been heard and properly addressed.
- People understand the value to be generated from the change.
- Change is being led by capable and competent leadership personnel and teams.
- The timing is right.

Organizational readiness for change is an essential assessment to be performed by leaders before any major change, such as venturing onto the path of OEMS, is considered.

6.13.4 Learning and Capability Development

For change to be sustainable, people must be trained and capable to support the required change. What this means is that before any change is proposed, people must be properly trained, competent, and assessed in skills so that they can perform at the new level of performance required by the organization. Where OEMS is concerned, training and ensuring workers are competent is an essential requirement for sustainable change. It is important to note that the level of training required may vary from expert competence to awareness, whereby those who lead the change must be experts, and supporters may only require awareness training.

6.13.5 Communication

In every organization, it is almost impossible to train every employee to the same level of competence. As a consequence, it is necessary to communicate proposed changes to all stakeholders who are likely to be affected by the change. Using the appropriate channel, change can be communicated to all stakeholder groups. Where OEMS is concerned, communication may require multiple channels of communication and should be repeated many times over so that the key messages of OEMS are transferred to all stakeholders.

6.13.6 Change Networks

Change networks may be required to generate and support OEMS change so that it may be sustainable. Assigning the tasks for ensuring sustainable change to a network has strong potential for ensuring sustainable change. Change networks must be treated much like any of the networks discussed in fundamental 7 and governed accordingly so that they may be ambassadors of change for the rest of the organization.

6.13.7 Creating a Culture of Learning and Continuous Improvements

Change is never static. Furthermore, OEMS is an evolving process. Organizations must continually evolve while considering the following initiatives:

- Adapt to business environment challenges
- Take advantage of technological improvements
- Adjust to personnel changes
- Take advantage of evolving and improved ways of doing the same thing
- Generate more efficient ways of doing things
- Learn from peers, industry, and incidents

Creating a culture of learning and continuous improvement is an essential requirement for OEMS and must be generated from demonstrated leadership behaviors.

6.14 Key Leadership Focus for Developing an OEMS

Is there a single leadership style or set of behaviors that will generate the best business performance or operational discipline and excellence? This is a difficult question to answer, but in the authors' view, the simple answer is no. Leadership, no doubt, is critical for developing operational discipline and operational excellence management systems. When properly analyzed, the authors recommend leaders should, at the very minimum, focus on the following to initiate and develop an OEMS.

- Sharing the vision:
  - Initiate senior leadership communication
  - Frequent town hall meetings
  - Facilitate engagement sessions
- Inspiring the hearts and minds of workers:
  - Demonstrate strong organizational values
  - Advocate respectful treatment
  - Develop cultural awareness, intelligence, and sensitivity
  - Create a trusting work environment
  - Motivate workers
- Managing change:
  - Lead and manage change
  - Identify and engage stakeholders
  - Utilize stakeholder impact assessments:
    - Business unit
    - Business area
    - Departmental
    - Individual
- Showing genuine care and empathy:
  - Change workloads
  - Consider different ways of doing things
  - Acknowledge personnel impact
- Promoting teamwork:
  - Create centers of excellence, networks, and communities of practices
  - Develop cultural awareness, intelligence, and sensitivity
  - Harness support and resourcing
  - Encourage stewardship
- Developing workers:
  - Direct
  - Support
  - Coach
- Delegate

References

Beyond Petroleum (BP). (2013). Our values and behaviors. Retrieved June 9, 2013, from http://www.bp.com/liveassets/bp_internet/bp_hungary/bp_hungary_european/STAGING/local_assets/downloads_pdfs/o/Our_values_and_behaviours.pdf.

Davis, J.H., Schoorman, F.D., Mayer, R.C., and Hwee Hoon, T. (2000). The trusted general manager and business unit performance: Empirical evidence of a competitive advantage. *Strategic Management Journal*, 21(5), 563–576. Retrieved November 6, 2006, from EBSCOhost database.

Hemdi, M.A., and Nasurdin, A.M. (2006). Predicting turnover intentions of hotel employees: The influence of employee development human resource management practices and trust in organization. *Gadjah Mada International Journal of Business*, 8(1), 42–64. Retrieved November 9, 2007, from EBSCOhost database.

Hopkins, S.M., and Weathington, B.L. (2006). The relationships between justice perceptions, trust and employee attitudes in a downsized organization. *Journal of Psychology*, 140(5), 477–498. Retrieved November 24, 2007, from EBSCOHost database.

Ismail, A.M., Reza, R., and Mahdi, S. (2012). Analysis of the relationship between cultural intelligence and transformational leadership (the case of managers at the trade office). *International Journal of Business and Social Science*, 3(14), 252–261. Retrieved June 9, 2013, from EBSCOhost database.

Ken Blanchard Group of Companies. (2010). Building trust: The critical link to a high-involvement, high-energy workplace begins with a common language. Ken Blanchard Group of Companies, Escondido, CA.

Lutchman, C. (2010). *Project execution: A practical approach to industrial and commercial project management*. Taylor & Francis, CRC Press, Boca Raton, FL.

Lutchman, C., Maharaj, R., and Ghanem, W. (2012). *Safety management: A comprehensive approach to developing a sustainable system*. Taylor & Francis, CRC Press, Boca Raton, FL.

Lutchman, C., Evans, D., Maharaj, R., and Sharma, R. (2013). *Process safety management: Leveraging networks and communities of practice for continuous improvement*. Taylor & Francis, CRC Press, Boca Raton, FL.

Ramlall, S. (2004). A review of employee motivation theories and their implications for employee retention within organizations. *Journal of American Academy of Business, Cambridge*, 5(1/2), 52–63. Retrieved September 30, 2006, from EBSCOhost database.

Wren, D.A. (1994). *The evolution of management thought*. 4th ed. John Wiley & Sons, New York.

Willink, T.E. (2009). Beyond transactional management: Transformational lessons for pharmaceutical sales and marketing managers. *Journal of Medical Marketing*, 9(2), 119–125. Retrieved June 9, 2013, from EBSCOhost database.

7

Fundamental 2: Identifying and Executing Applicable and Relevant Elements to Achieve Operational Excellence in Your Business

The required operational excellence management system (OEMS) elements of an organization may vary based on the following characteristics of the business:

- Risk and hazards exposures of the organization
- Type of process operations the business is involved in
- Size and scale of operation of the business
- Vision of the organization
- Number of changes occurring in the business
- Presence or absence of a working management system in the organization
- Organizational and functional structures of the organization

Nevertheless, when the organization decides to strive for an OEMS, it must first determine the elements of the management system that are applicable to its business.

In this chapter, the authors review all of the typical elements applied by organizations that are deemed necessary for operational discipline and operational excellence. In this chapter, the authors seek to categorize these elements into three major categories:

1. People elements
2. Processes and systems elements
3. Facilities and technology elements

Elements associated with each category shall be discussed in detail, and the requirements for each element will be identified such that practitioners can develop the applicable elements and their requirements based on applicable industry.

7.1 Elements Categorized into People, Processes and Systems, and Facilities and Technology

As discussed in Section I, common elements of an OEMS, as shown in Figure 3.3, were categorized into people elements, process and system elements, and facilities and technology elements. Categorization is not a precise science and may vary from organization to organization. However, the key requirement of an organization for success is to correctly identify which elements of the OEMS are required by the organization.

Table 7.1 provides an overview of the various elements of the OEMS, the aim of the elements, and the requirements for each. While not exhaustive, the list of requirements provided for each element can be expanded to provide greater rigor and more operational discipline within the organization.

TABLE 7.1

OEMS Elements, Aims, and Requirements

Element	Requirements
People Elements	
Element: Management, leadership, and organizational commitment and accountability **Aims:** • To ensure all levels of leadership and management demonstrate commitment to operational discipline and excellence • To define and steward appropriate accountability for operational excellence throughout the organization	• Shared vision for an OEMS. A focus on safety, reliability, and socially responsible operating principles. • Visible and demonstrated support for OEMS. Leaders visibly demonstrate commitment and personal accountability to improve operational integrity, performance, and discipline. Commitment is demonstrated by providing timely and required support and resources through visible participation in efforts to implement and improve the system. • Continuous ongoing communication on goals, performance, and achievements, while keeping interest and support high throughout the organization. • Integration of OEMS elements into business processes, practices, and decisions. • Pursue excellence in people, assets, and technology. • Improve self and followers on an ongoing and sustainable basis. • Clearly defined roles and responsibilities for all workers. • Ongoing performance management of followers, project teams, and contractors. • A focus on always doing the right thing. • A focus on doing it once only—the right way. • Through industry networking, industry recommended and best practices adopted by other leading organizations are considered for inclusion in the organizational OEMS plans and programs.

TABLE 7.1 (CONTINUED)

OEMS Elements, Aims, and Requirements

Element	Requirements
People Elements	
Element: Leadership and management review **Aims:** • To confirm that operational and management processes are implemented and assess whether they are working effectively • To measure progress and continually improve toward meeting operational integrity objectives, targets, and key performance indicators	• Formal processes for setting business targets and for stewarding the business performance. • Documented practices and data repositories for supporting transparency in business performance. • Leveraging leading and lagging indicators to establish rigorous targets and for stewarding business performance. • Ensuring visibility of metrics and performance so that all workers are aware of performance and required actions to improve and sustain performance. • Provide a framework for the reporting of performance data consistent with the corporate principles. • Provide and maintain a process for communicating to the workforce the existence, relevance, importance, and status of compliance programs and requirements. • Maintaining a process for demonstrating compliance to all legal and regulatory requirements.
Element: Legal requirements and compliance **Aims:** • To ensure *conformance* with internal policies, standards, procedures, and practices and *compliance* with all relevant legal requirements and government regulations • To conduct work in an ethically and socially responsible manner and in a sustainable way within communities and social environments while optimizing social value	• Development and sustainment of a process for identifying and applying all legal and regulatory requirements. • Maintain a process of managed change where operating outside of the legal and regulatory requirements. • Strive for performance and standards that exceed legal and regulatory requirements. • Maintain processes for demonstrating compliance with all legal and regulatory requirements. • Maintain a process for proactively identifying and addressing emerging regulatory requirements and for communicating these requirements to relevant stakeholders. • Maintain a process for evaluating compliance with regulatory requirements.
Element: Security management and emergency preparedness **Aims:** • To ensure preparedness for a security threat, business interruption, emergency, or crisis	• Develop and maintain a security management process that protects the organization's assets against threats of terrorism, unauthorized entry, and unplanned events. • Adopt and manage an emergency management process or model designed to address the risk exposures of the organization.

(continued)

TABLE 7.1 (CONTINUED)

OEMS Elements, Aims, and Requirements

<table>
<tr><th>Element</th><th>Requirements</th></tr>
<tr><td>People Elements<ul><li>To identify and verify all necessary actions to be taken to protect people, the environment, and the organization's assets and reputation in the event of a security threat, business interruption, emergency, or crisis</li><li>To ensure that supporting tools and processes are available to manage a security threat, business interruption, emergency, or crisis</li></ul></td><td><ul><li>Documented security and emergency management plans describe how emergencies will be managed. These plans shall comply with all industry recommended best practices.</li><li>Define roles and responsibilities for emergency response within the organization and provide adequate and designated resources to support an emergency response.</li><li>Ensure competent and capable personnel are assigned to emergency management roles, and these roles are periodically competency assured to ensure quality emergence responses when necessary.</li><li>Practice and perform drills on realistic and worst credible emergency scenarios to which the organization may be exposed.</li><li>Develop and maintain a process for capturing and sharing learning after each response or drill.</li><li>An emergency management procedure must address business continuity scenarios that include credible scenarios, such as pandemic conditions, labor unrest, and other realistic environmental and business conditions.</li><li>Business stakeholders, such as affiliates, subsidiaries, operating partners, and contractors, are encouraged to adopt practices consistent with the organization's expectations regarding security, emergency preparedness and response, and business continuity.</li></ul></td></tr>
<tr><td>Element: Qualification, orientation, and training

Aims:<ul><li>To ensure personnel possess the required competencies—knowledge, skills, abilities, and demonstrated behaviors to perform assigned work effectively, efficiently, and safely</li></ul></td><td><ul><li>Workers are properly oriented upon hiring and when assigned new roles. They are made aware of the inherent hazards and risks associated with the job.</li><li>The organization shall develop and maintain documented technical and professional competencies required to perform work.</li><li>No personnel shall be permitted to do work unless deemed competent to do so. Therefore, the organization must develop and maintain an up-to-date competency assurances process and records management process.</li><li>Competency assessment shall be conducted by qualified personnel against documented competency assurance requirements.</li></ul></td></tr>
</table>

TABLE 7.1 (CONTINUED)

OEMS Elements, Aims, and Requirements

Element	Requirements
People Elements	
	• When training is undertaken, learning objectives should be defined prior to delivery. Effective training shall adopt methodologies based on adult learning principles and shall include demonstration, engagement, job shadowing, mentoring, and other techniques to enhance learning. • Training materials and providers shall be evaluated for effectiveness, and training providers and training content may be updated to optimize training effectiveness. • Internal performance evaluation and management processes shall be implemented to monitor, measure, and evaluate conformance to requirements of policies, standards, business processes, and business and operating procedures. • Collective competencies of various work groups shall be maintained and assessed on an ongoing basis to ensure the integrity of work performed. For example, continuous operating facilities shall have balanced capabilities within shifts and work groups to ensure the reliability and integrity of operations.
Element: Contractor management **Aims:** • To ensure contractors and suppliers selected by the organization are best capable of performing work in a manner that is consistent and compatible with the requirements of the policies and business performance standards of the organization • To ensure contracted services and procured materials meet the requirements and expectations of standards of the organization	• Contractors and suppliers are appropriately prequalified and evaluated for the specific scope of work to ensure only the best performing contractors are selected. • The organization maintains a robust contractors and suppliers selection process, whereby selection is based on the ability to adequately demonstrate health, safety, quality, and management systems requirements that are closely aligned with those of the organization. • Contractors and suppliers are performance managed and are provided with timely feedback and corrective actions where required. • Contractor and supplier are *collaboratively audited* and gaps, improvement opportunities, and deficiencies are documented, prioritized for closure, and corrected in a time frame commensurate with risk exposures and value creation efforts of the organization. • Adequate performance management tools and processes are available to manage the ongoing performance of the contractor during the execution of the work.

(continued)

TABLE 7.1 (CONTINUED)

OEMS Elements, Aims, and Requirements

Element	Requirements
People Elements • To ensure that contractors and service providers selected by the organization are safe and effective with adequate standards and procedures that are aligned with the organizations	• A process is available to ensure the effective closeout of contracts. This process must also enable the organization to ensure poor performing contractors are excluded from future work.
Element: Event (incident) management and learning **Aims:** • To report and investigate all incidents • To ensure root causes of incidents are identified and appropriate corrective actions are applied to correct gaps and deficiencies in the management system of the organization • To capture learnings from incidents and to apply this knowledge to prevent recurrence of the same or similar incidents	• All stakeholders are required to report all incidents in the organization from which meaningful trends and analysis are generated, and from which knowledge designed to improve the overall performance of the organization is applied. • Incidents and near misses are regarded as learning opportunities for the organization and industry as a whole. • Effective root cause analysis (RCA) is conducted on all incidents. Corrective actions are generated and stewarded to closure. • The RCA process seeks to avoid blame assignment and finger pointing while focusing on identifying management system deficiencies to prevent recurrence of incidents. • The extent of the RCA conducted is dependent upon the severity of the incident or the potential severity from near misses. • Knowledge generated from incidents is shared within the organization. Where applicable, this knowledge is shared in an organized manner within the industry and across industries to prevent a repeat of the same or similar incidents.
Element: Management of personnel change (MOC-P) **Aims:** • To provide a disciplined approach for making required business changes that removes additional risk exposures to the organization from any changes made • To highlight and clearly define the organizational requirements for making changes	• Develop and maintain a list of critical roles in the organization such that when changes are made, additional risks are not brought upon the organization. • Develop and maintain a pool of competent resources for safety-critical roles such that attrition, turnover, and other unplanned changes do not introduce additional risks and hazards to the organization. • Ensure an adequate succession plan is maintained such that workers are not stagnated. They are inspired to remain with the organization despite being free to go and are provided adequate opportunities for growth and development.

TABLE 7.1 (CONTINUED)

OEMS Elements, Aims, and Requirements

Element	Requirements
People Elements	
• To provide a process for changes made by third-party stakeholders when working for the organization	• People changes in contracting and third-party providers are not made without consultation with the organization, and such changes are managed to avoid additional risk exposures to the organization.
Processes and Systems Elements	
Element: Establishing SMART (specific, measurable, achievable, realistic, and time bounded) goals, targets, and key performance indicators (KPIs) **Aims:** • To confirm that operational integrity management processes are implemented and assess whether they are working effectively • To measure progress and continually improve toward meeting operational integrity objectives, targets, and key performance indicators	• Clear goals and specific objectives are established for each element of the management system, and performance is evaluated against these goals and objectives. • Operational performance is periodically assessed and communicated to all employees, contractors, stakeholders, and the general public. • Leading and lagging indicators are established and stewarded to ensure the health and safety of personnel, environmental performance, process safety, and operational integrity on the assets. • Take corrective actions based on KPI stewardship such that operational discipline and integrity are optimized. • Report operating performance data and information in accordance with the organizational processes and requirements. • Require third-party stakeholders—contractors and subcontractors—to provide accurate and timely leading and lagging indicator information to create a complete organizational performance status.
Element: Ensuring the right organizational structure **Aims:** • To ensure the organization is structured to deliver on planned and approved business objectives in an efficient and effective manner while ensuring the use of competent people and adequate resources.	• An organizational structure with defined roles and responsibilities that promote authorities, accountabilities, and responsibilities is available and accessible by workers. • Provides clear line of reporting with a manageable span of control for each leader. • The organizational chart is maintained as current, and personnel are evaluated periodically to ensure the strategic goals, operational integrity, and operational discipline of the organization are maintained. • Role description, training requirements, and responsibilities are attached to each position in the organizational structure.

(continued)

TABLE 7.1 (CONTINUED)

OEMS Elements, Aims, and Requirements

Element	Requirements
Processes and Systems Elements	
• To ensure responsibilities and authorities are clearly defined and unambiguous; leaders can therefore be held accountable for performance	• Defines roles that are permanent and those that are temporary. • Identifies roles that may be held by contractors. • Shows integrated relationships with third-party (contractor) organizations as applicable.
Element: Risk management **Aims:** • To provide a systematic framework for operational risk identification, mitigation, and management • To prevent incidents by identifying and minimizing workplace hazards and personal health risks	• Risk identification and mitigation process. • A standardized risk matrix that is consistently applied for all risk exposures of the organization. • Create and steward a risk register that identifies the risk, determines the mitigation actions needed, assigns the risk management responsibilities, and updates the corrective actions taken. • Documents the risks identified and the mitigation actions taken to reduce risks to as low as reasonably practicable (ALARP) levels. • Work processes that leverage layers of protection and the concepts of the Swiss cheese model, such that engineered solutions for risk reduction is the most preferred and applied. • Work is controlled such that risk is mitigated to ALARP levels and appropriate personnel protective equipment (PPE) becomes the last line of defense during the execution of the work. • Field-level risk assessments and field-level hazard assessments are done prior to the execution of any work.
Element: Environmental management system **Aims:** • To demonstrate the right level of social responsibility by minimizing environmental impacts from the business operations • To create a sustainable environmental legacy whereby the impact of the business operation is negligible to the environment	• Implement practices and procedures that minimize, prevent, and mitigate environmental impact from operations. • Develop environmental monitoring programs that provide early indication of adverse environmental impacts such that preventative measures can be activated. • Environmental impacts are monitored and reported to demonstrate conformance with internal policies and procedures and compliance with local, national, and international laws and regulations as applicable.

TABLE 7.1 (CONTINUED)

OEMS Elements, Aims, and Requirements

Element	Requirements
Processes and Systems Elements	

Element: Management of engineered and nonengineered change
 	
Aims:
• To provide a disciplined approach for making required process and business changes that removes additional operational risk exposures to the organization from any changes made
• To highlight and clearly define the organizational requirements for making changes
• To provide a process for changes made by third-party stakeholders when working for the organization
	• A formal process for managing change is required that addresses the following:
	  • Definition of a change
	  • Process for initiating change
	  • Documentation requirements for change
	  • Subject matter experts' requirements to initiate change
	  • Authority and authorization process for activating change
	  • Risk assessment of the change
	  • Stakeholder impact of the proposed change
	• Responsibilities for executing change. Assigned responsibilities for approving, implementing, and risk mitigating actions and verifying completeness and effectiveness of the change are documented.
	• A formal process for establishing impact of the change and for communicating the change impact to all stakeholders.
	• A documented action items management log is maintained and stewarded to ensure all required actions to ensure successful implementation of the change are achieved.
	• A formal evaluation of the effectiveness of the change is performed. This evaluation is designed to determine if the objectives of the change were accomplished and if unintended risks and hazards are introduced into the new status.

Element: Communication, stakeholder engagement, and stakeholder management **Aims:** • To provide a sustainable framework for stakeholder communication such that reputation and goodwill are maintained • To provide a sustainable framework for engagement of stakeholders	• Processes for internal communication are developed and implemented such that workers are engaged and involved in business decision making. • Processes for external communication regarding business performance, emergency management, and public relations management are developed and maintained. • External communication processes must have the appropriate levels of approval for sharing information outside the organization. • A feedback and communications evaluation process is required to assess the impact of all internal and external communication. • A documented stakeholder engagement process for involving stakeholders in the business process is required.

(continued)

TABLE 7.1 (CONTINUED)

OEMS Elements, Aims, and Requirements

Element	Requirements
Processes and Systems Elements	
	• A process shall be maintained to ensure that all workers or authorities required to represent the organization at external forums and communication sessions are trained and competent in doing so. • Ensure communities and stakeholders impacted by the business processes and operations are consulted and communicated with on a sustainable frequency via a dedicated and assigned authority. • Develop and maintain a process for maintaining strong stakeholder relationships. • Consider external stakeholders in the change impact assessment process when operating changes are made.

Element: Prestart-up safety reviews
 	
Aims:
• To have a rigorous prestart-up review process that ensures overlooked hazardous conditions are avoided
	• Develop and maintain a standardized prestart-up safety review process that addresses unsafe conditions and hazards that may be present after the following:
	  • Construction of a new facility
	  • Turnaround after major maintenance
	  • Start-up after engineered and nonengineered changes have been made
	• Maintain a documented process for stewarding gaps identified to closure.

Element: Operations integrity audits **Aims:** • To proactively identify systematic and single-point gaps in the management system and close them before incidents can occur • To demonstrate good business practices in the stewardship of the organization	• Develop and implement a risk-based self-assessment program to assess conformance to the management systems requirements and compliance to legal and regulatory requirements. • Develop and steward a documented prioritized corrective actions plan for addressing gaps identified during self-assessments. • Develop a stewardship process for tracking and verifying that corrective actions are implemented and are working as per management systems requirements. • Develop and implement a risk-based *collaborative* contractor audit program to assess contractors' conformance to and alignment with the organization's management systems requirements and compliance to legal and regulatory requirements. • Develop and steward a documented prioritized corrective actions plan for addressing gaps identified during contractor audits. • Develop a stewardship process for tracking and verifying corrective actions implemented by contractors who are working as per management systems requirements.

TABLE 7.1 (CONTINUED)

OEMS Elements, Aims, and Requirements

Element	Requirements
Facilities and Technology Elements	
Element: Process safety information **Aims:** • To ensure the safe management of important information and documentation required to safely operate and manage the business • To provide easy and controlled access of relevant information by workers required to operate the business safely and reliably • To provide a framework for the sharing of industry best practices and safety-critical knowledge	• Develop, implement, and maintain an information management process for the management of data, reports, audio, video, studies, and other relevant information and make available and accessible to users at the right time. • The document management system should allow for the controlled access and use of information, for example, read/write access and privileged and confidential access. • Access to procedures, policies, standards and records and documentation relating to operations, maintenance, inspections, and facility changes is maintained, and these are updated as per the organization's frequency and processes, and are auditable. • Data, documents, and information shall be managed in such a way that required backups and redundancy are maintained. • Data, documents, and information shall be managed in such a way that outdated and obsolete processes, procedures, and information are removed from circulation to avoid errors that may bring about harm to people or damage to the environment or the business assets.
Element: Process hazards analysis (process safety) **Aims:** To ensure the integrity of hazardous operations by the application of sound process safety management processes and principles	• Develop processes that align with process safety management (PSM) requirements (where not regulatory) and implement PSM requirements where regulated to ensure risks and hazards are systematically removed from the process by leveraging people, processes and systems, and facilities and technology risk reduction techniques and measures. • Determine worst credible major incident scenarios and design engineered solutions to protect people, the environment, and assets. • Develop processes that leverage both internal and external expertise in risk reduction strategies.
Element: Physical assets systems integrity and reliability • Operations and maintenance controls • Quality assurance • Mechanical integrity	• Develop and maintain a formal process for identifying and documenting potential risks and hazards and the operational and maintenance controls to control these hazards consistent with layers of protection principles. Design out hazards, guards and barriers, administrative controls, and PPE.

(continued)

TABLE 7.1 (CONTINUED)

OEMS Elements, Aims, and Requirements

Element	Requirements
Facilities and Technology Elements	
Aims: • To ensure operations are maintained within the safe operating design limits of the facility and that preventive maintenance is conducted as per manufacturers and design requirements • To ensure that work performed by third-party providers is done under strict quality management standards such that the integrity of the operating assets is in no way compromised	• Develop and maintain processes to ensure operational and maintenance controls are implemented and evaluated for effectiveness. • A process is implemented to ensure critical spares, structures, equipment, and protective devices are available when needed. • Develop and maintain processes for testing and validating protective equipment. • Develop and implement processes and standards that guide the work of third-party providers such that the operating envelope of the asset is not compromised when work is completed—quality assurance standard. • Develop and implement systems and processes to monitor, report, manage, maintain, inspect, and test backlogs so that augmentation can occur, particularly on safety-critical systems.
Element: Operating procedures and safe work practices **Aims:** • To provide documented instructions and guidance (as applicable) to the workforce on how to safely and efficiently perform nonroutine work • To provide a standardized work permitting process that identifies and mitigates task-specific hazards before any work can be done	• Develop and implement a process whereby current and up-to-date procedures are available and accessible to all users (employees and contractors performing work on behalf of the organization) as required. • Ensure contractors follow the organization's procedures for performing the work unless the contractor's procedures are of equivalent or higher standards. • Develop and implement a process for ensuring that procedures are updated at an approved frequency and immediately when found to be deficient in their application. • Develop and maintain a process that ensures all users are trained on updated procedures prior to their use of any updated procedure. • Ensure procedures are written with the users in mind and provide answers to the questions presented by users. Adopt *usability mapping principles* in the creation of procedures and guidance documents. • Ensure a safe work permitting process is implemented for all nonroutine work that is designed to protect all workers during the execution of work. • Ensure the safe work permitting process is applicable across all work groups. • Ensure an auditing process is implemented to ensure the validity of the safe work permitting process.

7.2 Conclusion

The requirements for each element will be determined by the respective organization. Organizations may choose to expand on the list of requirements provided in Table 7.1, or repackage accordingly based on the number of elements the organization chooses to establish and steward. Requirements defined in Table 7.1 may be broken out further to ensure simplicity, standardization, and more effective control within the organization.

References

Lutchman, C., Evans, D., Maharaj, R., and Sharma, R. (2013). *Process safety management: Leveraging networks and communities of practice for continuous improvement*. Taylor & Francis, CRC Press, Boca Raton, FL.

8

Fundamental 3: Establishing the Baseline

Knowing where the organization performs relative to the requirements of each element identified for the management system is an important step in creating an operational excellence management system (OEMS). This step is essentially establishing the baseline of existing organizational performance for each element. Knowing the baseline provides opportunities for the organization to strategically and purposefully achieve organizational requirements for each element.

An objective scoring process is required to help the organization determine the baseline for each element. The scoring process starts at the element requirements level. Consistent with the requirements identified in Table 7.1, each requirement of the element is assessed relative to the scoring criteria defined in Table 8.1. This analysis provides a numeric evaluation of the OEMS status of each element. Once all of the elements of the business unit or organization are assessed, the OEMS status of the business unit or organization is established.

Figures 8.1 and 8.2 provide a graphical presentation of a likely baseline assessment outcome for a typical element and business unit or organization, respectively. Figure 8.1 provides an assessment of the subelements for the element (event incident management and learning) based on the scoring rubric defined in Table 8.1. When all elements of the management system are similarly assessed, the baseline assessment for a business unit of the organization is completed. Typically the baseline assessment is done as follows:

1. Assessment of subelements.
2. The average score of the subelements produces the score of the element.
3. The average score of all elements of the management system generates the business unit or organizational score.

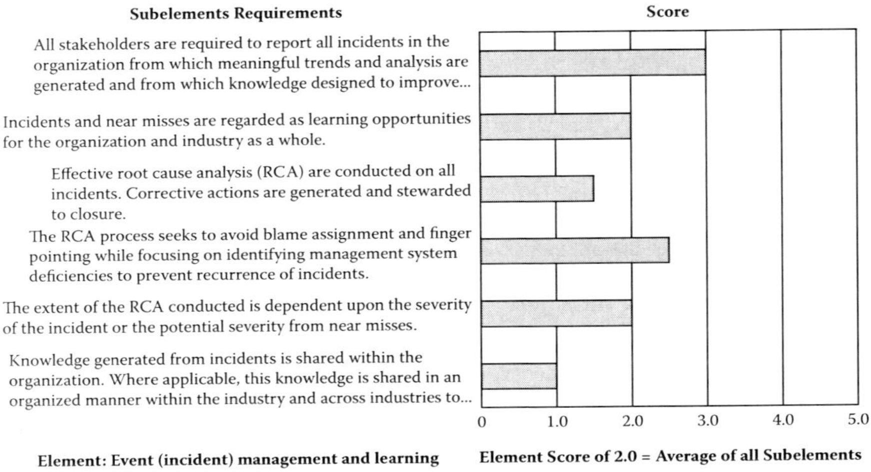

FIGURE 8.1
OEMS baseline assessment for an element of the management system (shown by the asterisk in Figure 8.2).

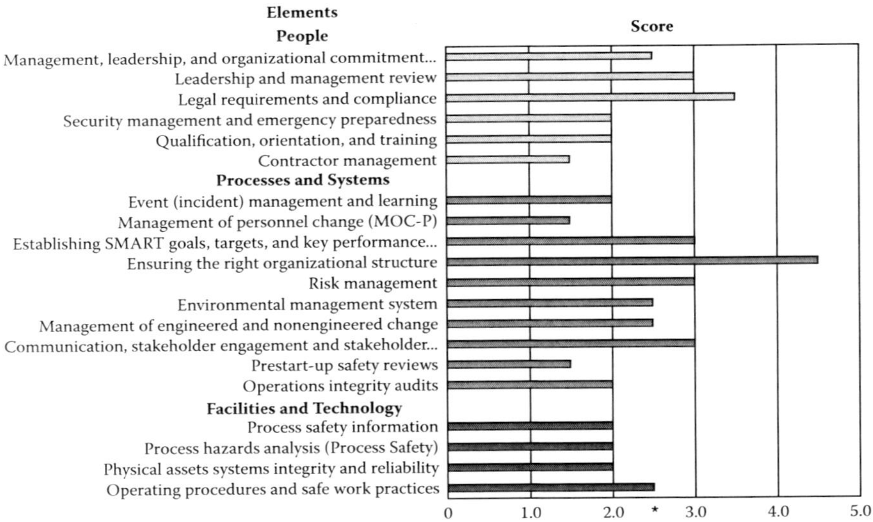

FIGURE 8.2
OEMS baseline assessment for a business unit or organization.

TABLE 8.1

Organizational Capabilities for Establishing the Baseline for Each Element

Organizational Capabilities	OEMS Status	Score	Organizational Performance
• Informal • Inconsistent • Ad hoc • Uncoordinated • Noncompliant • People dependent • Ineffective	Regressive	1.0	Chaotic and firefighting modes of operation
• Emerging • Partial implementation • Minimum compliance • Firefighting • Rework • Repeat incident	Reactive	2.0	Inefficient and disorganized
• Implemented • Standardized • Repeatable • Consistent • Compliant • Managed risks and hazards • Learning	Planned	3.0	Efficient and organized
• Integrated • Predictable • Ownership • Beyond compliance • Continuous improvement • Risk elimination • Quantitatively managed • Learning within organization	Proactive	4.0	Disciplined and excellent
• Leading • Strategic • Culture driven • World class • Optimized • Organizational shared learning • Industry best practice • Innovator	Excellence	5.0	Operational excellence—world class

8.1 Preparing for the Baseline Assessment

Adequate preparation is required for success in establishing the OEMS baseline. Among the critical preparation requirements are the following:

1. Providing a framework for a standardized and consistent interpretation of the requirements for each subelement
2. Identifying and involving element champions/subject matter experts (SMEs) for performing baseline assessment work

Generally, the OEMS baseline is best determined by the business unit personnel and verified by an objective and impartial team of management systems auditors. When the baseline assessment is performed by the business unit experts, we term the process baseline self-assessment (BSA).

8.1.1 Selecting Element Champions/SMEs

To complete the BSA, element champions who possess the following skills and attributes are required:

- Subject matter expertise for the particular element
- Objectivity and impartiality—unbiased analysis
- Auditing skills
- Organizing skills
- Nonauthoritative leadership skills
- Ability to put people at ease—demonstrate genuine empathy and care for people
- Ability to conduct noninvasive probing
- Respected in the organization

Once element champions are selected for each of the management system elements, their job then transitions to determining the state of compliance/conformance with the requirements of each subelement of the particular element by the business unit. The level of compliance/conformance eventually produces the element baseline assessment as shown in Figure 8.1.

8.1.2 Standardized and Consistent Interpretation

Where multiple business units are present in the organization, it is important that BSAs conducted by the different business units are done with strong consistency. As a consequence, therefore, interpretation guidelines for the requirements of each subelement are essential to facilitate the process. Consistent interpretation of the requirements by element champions across multiple business units allows for apples-to-apples comparisons.

In this way, gaps identified and best practices opportunities are real and consistent. Consistent gaps provide opportunities for the organization to establish corporate solutions for common gaps. Consistent proactive and excellence performance allows for transfer of best practices from high-performing business units to reactive and regressive-performing business units.

Standardized and consistent interpretations of subelement requirements therefore require interpretations that are characterized by the following:

- Clear and concise
- Simple and easy to follow
- Unambiguous
- Provide quantitative measures for measurable parameters
- Objective

For best consistency during the BSA, element champions should be brought together from all business units and appropriately trained in the interpretation guidelines.

8.2 Conducting a Baseline Assessment

Conducting the baseline assessment of a business unit requires the combined and coordinated efforts of two groups, each performing a specifically assigned task. These groups are aligned on the common goal of providing an adequate representation of the management system status of the business unit or organization relative to its *planned* state. These groups are distinguished as follows:

- Group 1: Business unit-based element champions—perform the baseline self-assessment.
- Group 2: Expert management system auditors (verification team)—verify the scoring of group 1.

When conducting the BSA, element champions are required to provide compelling evidence that the requirements of each subelement meet the following evaluation criteria:

1. The requirements are adequately supported with documented formal processes, procedures, standards, and practices.
2. People are trained and competency assured in required procedures, processes, standards, and practices.

3. Processes are repeatable and sustainable.
4. Workers demonstrate consistent use of required procedures, processes, standards, and practices.
5. Assurance processes are in place to verify conformance and compliance.

When the requirements of each subelement are assessed against these evaluation criteria, a score can be applied as shown in Figure 8.1.

8.3 Verifying the Baseline Assessment

Once the self-assessment BSA is completed by the business unit element champions, the next step is to verify the subelements/element scoring. Verification is done by an expert management system audit team whose role is to examine the evidence provided by the element champion in support of allocated scores. The element champion participates in an interview process with the verification team where the element champion provides adequate justification for the allocated score for each subelement.

The verification team looks for the following as supporting evidence for scores allocated:

- Documented processes are updated, relevant, approved, and accessible.
- Accountabilities are assigned, and those who are accountable are aware of these accountabilities.
- Personnel required to use these processes are trained and competent and know how to use the processes. They are also able to locate and access all required and relevant procedures, standards, policies, and processes as and when required.
- There is an assurance process to ensure the quality of procedures, standards, policies, and processes are fit for purpose and continuously improved.
- The verification team tests the scoring of the element champion against the organizational capabilities identified in Table 8.1.

Occasionally, the verification team may choose to downgrade a score where insufficient evidence was provided to support the allocated score. On the other hand, the verification team may upgrade a score when the element champion is too harsh in scoring or in situations where other business units may have a similar process that may have been scored higher with similar organizational capabilities.

8.4 Conclusion

Establishing the baseline of the organization's management system allows the organization to prioritize opportunities to develop its processes and systems in an organized manner to achieve operation discipline and excellence. The key benefit of the baseline is knowing where the organization sits relative to where it wants to be from a capability perspective on each of its management systems elements.

Determining the baseline is among the key items in taking the organization from ordinary to extraordinary, from good to great, and along the path of discipline and excellence. Organizations now have the opportunity to develop strategies for closing gaps identified and for leveraging best practices identified in some parts of the business across the entire organization.

9

Fundamental 4: Plan–Do–Check–Act (PDCA)

The authors propose following the PDCA model as the fourth fundamental of an operational excellence management system (OEMS). Once the baseline self-assessment (BSA) is established, the organization should apply the following approach to closing gaps identified in the management system for each element:

- Prioritize gaps identified for closing them and leveraging best practices.
- Follow the PDCA model for closing the gaps and leveraging best practices.

In this section, the authors discuss a gap prioritization process and how the organization closes these gaps.

Figures 9.1 and 9.2 identify gaps within the element and the management system, respectively. The challenge for the organization is how to prioritize the gaps identified within each element and within the management system so that the organization is not overloaded and stretched to its limit. An overload as such can potentially lead to adverse impacts and a lack of attention on other subelements and elements of the management system.

It is important to remember that closing gaps identified within an element and within the management system depends on people resources within the organization or business unit who are also tasked with their day-to-day responsibilities. Since most organizations are prepared to close management system gaps with internal resources, careful attention is required for prioritizing and following the PDCA model to ensure people are not overworked or pushed to the point where they contemplate turnover within the organization, leaving the organization or *quitting and staying*. All of the aforementioned scenarios are extremely detrimental to the organization.

9.1 Prioritizing Gaps and Best Practices Identified for Closure

How do organizations prioritize work to be done? In most instances there are two key drivers:

1. Risk exposures reduction
2. Financial benefits

More recently, value maximization has been introduced into the decision prioritization process. In this section, the authors introduce the concept of

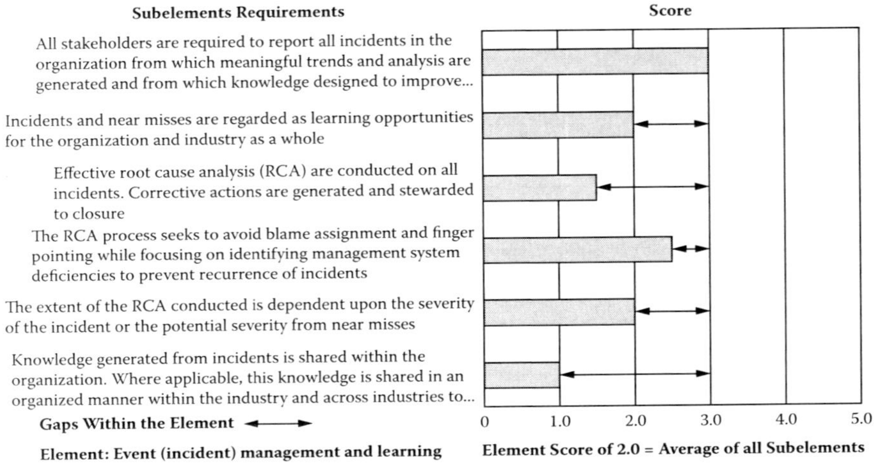

FIGURE 9.1
Gaps identified within the element when the BSA is completed.

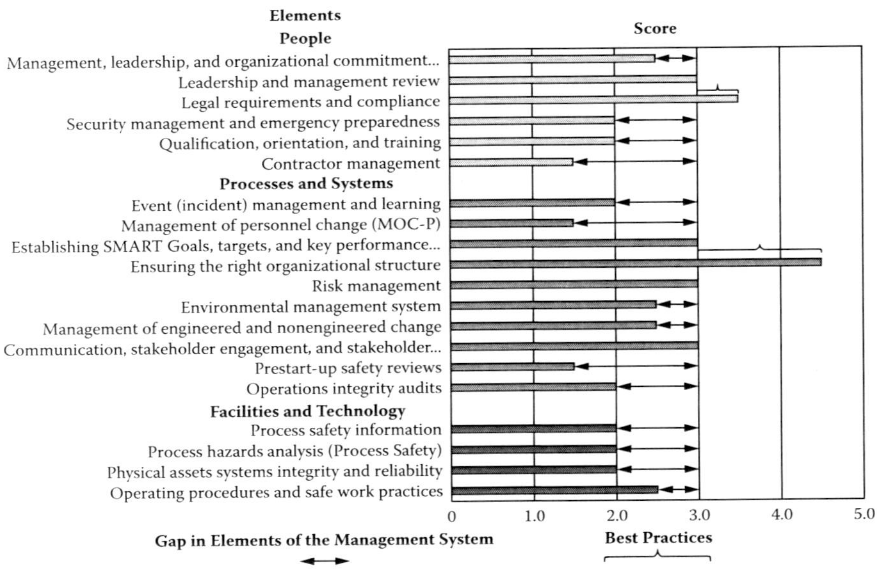

FIGURE 9.2
Management system gaps and potential best practices.

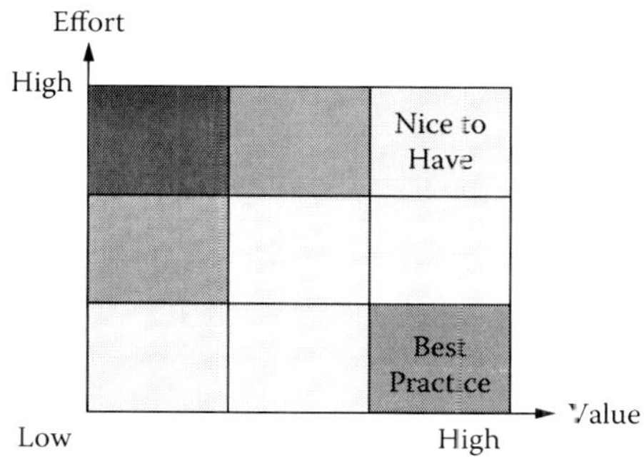

FIGURE 9.3
Simple value–effort matrix for prioritizing work. (From Safety Erudite, Fundamentals of an Operationally Excellent Management System Workshop, 2014, retrieved April 29, 2014, from www.safetyerudite.com.)

the opportunity matrix for decision making and prioritization of work.

In their simplest forms, however, decision making and prioritization are the outcomes of the *value–effort* relationship as shown in Figure 9.3. Work that generates high value with low efforts is prioritized ahead of that where the efforts required are higher and the value generated is low. At the extreme of this relationship are the following:

- Nice to have: High effort/high value
- Best practices: Low effort/high value

The priority matrix shown in Figure 9.4 is a bit more complex and objective than the value–effort relationship process for an OEMS element on contractor safety management. Here, the focus is on prioritizing work based on the inherent complexity of the work.

In this instance, a score is allocated based on the value created from work required and the complexity associated with completing the work. The complexity is determined by subject matter experts who can objectively assign a numerical value to the complexity of executing the particular work. While not an exact science, among the key factors determining complexity are the following:

- Number of stakeholders involved and impacted
- Impact on stakeholders
- Resources required for completing the work—people, finances, time, materials, and coordination
- Timing of execution, e.g., during weekends or night periods when there is limited personnel on site

This model may assist in prioritizing work both within an element and across elements of the management system.

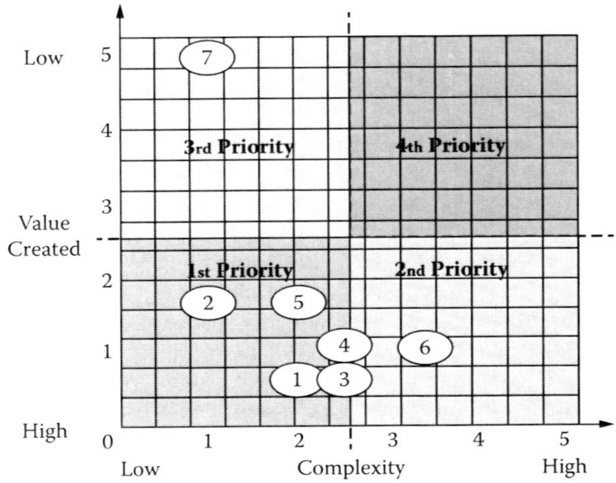

Priority	Opportunity	Complexity 1=Low 5=High	Value Created 1=High 5=Low
1	Get prequalification rolled out right	2.0	0.5
2	Align contract language with PSM/OMS requirements	1.0	1.5
3	Standardized Computer Based Contractor Orientation (3 levels)	2.5	0.5
4	Simplify CM101 and make BU/BA specific (workshop with tools)	2.5	1.0
5	CM 101 support to BU in rollout of Contractor Safety	2.0	2.0
6	Provide access to organization's work related information (procedures, policies, processes) to all contractors	3.5	1.0
7	Update the Contractor Safety Standard based on improvement opportunities identified	1.0	5.0

FIGURE 9.4
Opportunity matrix for prioritizing work for an OEMS element. (From Safety Erudite, Fundamentals of an Operationally Excellent Management System Workshop, 2014, retrieved April 29, 2014, from www.safetyerudite.com.)

9.2 Following the PDCA Model for Closing Gaps and Leveraging Best Practices

The PDCA model is perhaps the most widely applied model for managing work. First introduced by Deming in the 1950s (Moen and Norman, 2006), the PDCA model has found its way into almost all business processes and activities. It is the simplest model for managing any work. The

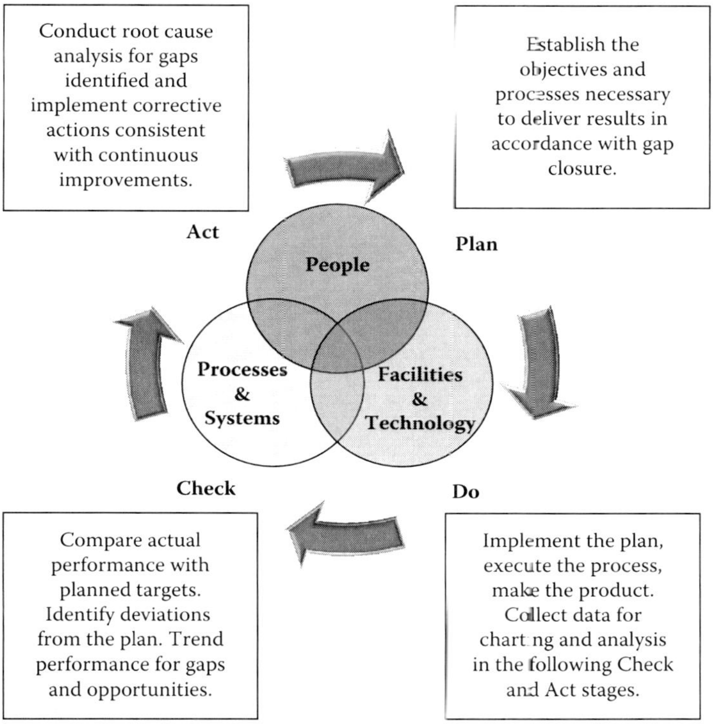

FIGURE 9.5
Application of the PDCA model. (From Safety Erudite, Fundamentals of an Operationally Excellent Management System Workshop, 2014, retrieved April 29, 2014, from www.safetyerudite.com.)

PDCA model is widely used in organizational environmental health and safety (EHS) processes and management systems and is extremely useful in closing managing work related to closing gaps in an element or management system.

Shown in Figure 9.5 is a graphic application of the PDCA model for closing gaps in an OEMS. The details of each step of the model are discussed, highlighting the key requirements for each step.

9.2.1 Plan (Planning)

Planning involves teamwork. Developing the plan for closing gaps for each subelement requires the following:

- Creating a team of workers led by a good team leader
- Having a clearly defined scope of work, often defined in a charter

- Defining the resources required to perform the work
- Developing the schedule complete with milestone activities to be completed
- Developing the operational controls and targets for stewarding the work
- Conducting a stakeholder impact assessment to understand and mitigate risks to successful implementation

Once a plan has been developed, the next step is to execute the plan according to the requirements of the plan.

9.2.2 Do (Doing)

The do (doing) stage of the PDCA model requires a clear focus on who, what, when, and where. The doing stage requires the following:

- Mobilization of resources
- Trained and competent people to do the work
- Simple procedures so that workers can complete all assigned work
- Data collection and key performance indicator (KPI) stewardship to determine progress relative to goals
- Mitigating risks and hazards determined from the stakeholder impact assessment

The doing stage is perhaps the most complex, and like the execution stage of a project, solutions and the final products take shape here for closing the prioritized gaps or rolling out best practices.

9.2.3 Check (Checking)

Once the plan has been executed and has been working for some time, the next stage requires checking to verify the gap has been closed or the best practice has been generating benefits as planned. Checking can take the form of a simple trending analysis or process to determine how the business unit or organization is performing planned targets. Additionally, checking can be more elaborate in auditing and scoring to assess performance relative to the BSA scoring.

Checking sets the framework for corrective actions management. Therefore, during this stage, early warning is crucial for success. The following are requirements for the checking stage:

- Verification against approved performance targets.
- Consistent approaches to checking and verification—standardized methods and tools to remove subjectivity.

- Simple processes that allow for appropriate checking frequencies.
- Stakeholder awareness of what is being checked.
- Checking lends itself to trends that may easily show positive and negative outcomes and excursions outside of acceptable ranges.
- Checking processes provide compelling evidence for corrective actions.

9.2.4 Act (Acting)

Based on findings from the checking stage, corrective actions management is initiated. Acting requires the application of root cause analysis (RCA) techniques to understand the prevalence of gaps at the root level. This ensures that corrective actions can be activated to guarantee performance improvements to close gaps and remove systematic problems that prevent the business unit or organization from achieving its goals. The following are requirements for the acting stage:

- A corrective management process that includes the following:
  - What is to be done
  - To whom the task is assigned
  - By when the task should be completed
  - What actions were taken
  - Who approves that the corrective actions were completed and executed
  - Any further follow-up requirements
- Stakeholder impact assessment on corrective actions to be executed

9.3 Conclusion

The BSA is perhaps one of the more important steps in developing an OEMS. At both the business unit and organizational levels, knowing where gaps in the management system's best practices may exist provides the organization opportunities to systematically remove gaps in the management system, while at the same time leveraging best practice opportunities.

Once gaps and best practices are identified, the key to success in closing these gaps and leveraging best practice opportunities lies in following the management principles of the PDCA model. This framework provides for organized and disciplined actions involved in closing gaps and leveraging best practice opportunities. Prioritization of gap closure work helps the organization to manage risk exposures and benefit from value maximization opportunities.

It must be noted, however, that not all gaps that have been identified have to be closed. A business unit or the organization may choose to deliberately maintain a gap between the current state and requirements of the standard. In such instances, the business unit will evaluate the risk exposure and decide that there are other priorities available for the organization. The decision, however, is informed and accepted by senior leaders and decision makers within the organization.

References

Moen, R., and Norman, C. (2006). Evolution of the PDCA cycle. Retrieved April 27, 2014, from http://pkpinc.com/files/NA01MoenNormanFullpaper.pdf.

Safety Erudite. (2014). Fundamentals of an operationally excellent management system workshop. Retrieved April 29, 2014, from www.safetyerudite.com.

10

Fundamental 5: Auditing for Compliance and Conformance

Audits today are slowly shifting from the perceptions of a policing function to a collaborative, proactive approach to improving organizational performance. The benefits of collaborative auditing are tremendous, and the level of cooperation received from business unit leaders is much stronger when the audit is a collaborative approach to improving the business unit performance. However, the challenges from a historically rooted stigma that audits are performed only when things are bad or when a business is underperforming continues to weigh heavily on the minds of leaders of the audited entity.

This fundamental focuses on the various aspects that relate to auditing for compliance and conformance. Auditing has become a very important process used by organizations to maintain an effective watch over business performance. Moreover, audits have shifted in focus to environmental health and safety (EHS) risk management within an organization. From an OEMS perspective, we regard auditing as a critical component for achieving operations discipline.

To provide you with a better understanding of auditing and its values, we attempt to distinguish among the various types of audits. Care must be taken to avoid confusing audits with inspections and the different types of audits employed within the industry. Also, what is important is demonstrating key result areas in audits and comparing strengths between different types of audit tools and processes. Some examples are discussed, and through implementation experience, a critique of different tools is provided. The reality is that all audit types provide value and possess strengths, as they were designed to address specific requirements.

10.1 Audits as an Organizational EHS Risk Management Tool

Almost all organizations that are governed by an EHS policy have a set of governing principles with objectives, elements, and expectations, which are holistically called an EHS management system. Today, the use of integrated EHS management systems with elements and expectations is especially important in organizations that have more than one operation or operate in more than one location or are geographically dispersed. EHS management systems often integrate other related elements in quality and security leading to health, safety, security, environment, and quality (HSSEQ) and health, safety, security, and environment (HSSE) management systems.

Many of these management systems adopt elements and expectations from or are otherwise based more on international best practice frameworks as prescribed in codes of practice and recommended guidelines, such as those from:

- Oil and Gas Producers (OGP)
- American Petroleum Institute (API)
- International Organization for Standardization (ISO)
- Local and international health and safety regulatory bodies, e.g., occupational health and safety (OH&S) and Occupational Safety and Health Administration (OSHA)

In some areas of the world and jurisdictions, there is also an expectation from governments/regulators that oil and gas companies, for example, have very high standards as they operate in a very hazardous environment. Many organizations have developed over the years with strict regimes of internal and external audits and compliance reviews to help them self-regulate and govern.

Technical audits have played an important developmental role in terms of their value, which is now comparable with operational audits. Increasingly, leaders have started integrating EHS into the following audits:

- Asset and operational integrity audits
- Business excellence and operational excellence audits

Some organizations have been driven to a more validation-type system auditing related to the ISO 90001, 14001, and OHSAS 18001 certifications, which focus more on the management system review and compliance, but also require auditing. Auditing is now prescribed as a key tool in general.

There are many other types of tools used in conjunction, which are not audits, but rather more types of risk assessments. These include hazard and operability (HAZOP) studies, hazard analysis (HAZAN) studies, job safety analysis (JSA), and process hazard analysis (PHA) type assessments, which are also very much used in projects and start-up/commissioning. It must be appreciated that major risk assessment reviews now take place at the plant level, driven mainly to prevent catastrophic failures and incidents. These have helped improve the safety record in the industry and perhaps remain one of the strongest elements of process safety management (PSM).

10.2 Differentiating between Audits and Inspections

There are fundamental differences between audits and inspections. The authors offer three definitions below that are important for any practitioner to better understand the difference between audits and inspections.

10.2.1 Audits

Audits are systematic and documented verification processes used to evaluate the success of an EHS management system in fulfilling the objectives and targets of the EHS policy, and its compliance with statutory regulations. Audits are generally conducted by trained, qualified, and competent auditors and subject matter experts (SMEs) who are generally tasked with comparing the organization's compliance or conformance to regulatory or internal requirements, respectively. There are generally three types of audits performed by the organization:

- Self-assessments: Audit performed by the organization itself.
- Second-party audits: The operation is audited by another part of the business.
- Third-party audits: External auditors are brought in to evaluate the degree of compliance or conformance of the organization.

Audits are generally supported with a comprehensive audit report that highlights gaps and opportunities for improvements, as well as best practices opportunities that can be shared across the organization. Approved audit reports adopt a formal approach to gap closure, whereby findings are categorized based on risk exposures and resources, assigned, and stewarded to closure. Audits focus on management system deficiencies and opportunities.

10.2.2 Internal Audits

Internal audits are audits performed by the company itself or by a third-party on behalf of the company as an internal review of the system's functionality. A third-party internal audit can be considered an external audit if the quality of the audit is approved by the corporation. When an audit is conducted by SMEs from another business unit of the corporation, it is called a second-party audit. It is recommended that internal resources are developed to undertake these audits, as they add greater value in terms of the system and people.

10.2.3 Inspections

These can be considered as routine checks done at certain frequencies to assure the operability of equipment and associated systems and worksites. This is generally guided by a (standardized) checklist or, for more extensive inspections, a regime or protocol. These are highly directed by the system itself and are very standardized. The key difference between an audit and an inspection is the focus: audits are focused on management system functionality, whereas inspections are focused on the physical condition at the worksite.

10.3 Purpose of Audits

Audits can serve many purposes within an organization and many purposes at the same time. The success of an audit also depends heavily on the competence of the auditor. A trained and competent auditor is someone who has a relatively good understanding of the operations of the organization that he or she is auditing. Furthermore, good auditors have strong analytical capabilities and transformational leadership traits that help put people at ease and encourage them to share and talk about their operations.

Audits can be used for the purposes in the below subsections.

10.3.1 A Management Tool

Audits are an extremely powerful management tool. When used correctly, audits provide a powerful opportunity to proactively identify and address issues and gaps within the management system. Audits may also be applied as a diagnostic tool that can help evaluate where improvements can be made to reduce risks.

10.3.2 A Verification of Compliance and Alignment

Audits can be used to check business performance against requirements of the EHS policies and procedures, PSM, and management systems of a particular organization to determine the extent of compliance. Such audits provide a sense of vulnerability of the organizations based on the regulatory and legal gaps identified.

10.3.3 Self-Assessment

Self-assessment is a process whereby leaders and line managers perform audits on their own operations to gauge their level of compliance and conformance to the business requirements. Self-assessments are very much a management tool and bring great value to the organization for prioritizing work and removing vulnerabilities from the business operations.

10.3.4 Benchmarking Performance

Audits are often used as a means of benchmarking organizational performance against a prescriptive standard or set of policies. Where a more generic standard is used, such as an international management system code, organizations can use this to benchmark themselves against other organizations in the same industry.

10.3.5 Evidence of Conformance and Improvements

Audits are generally used to demonstrate to stakeholders that sufficient and systematic efforts have been undertaken to ensure EHS and management systems compliance and conformance efforts are being delivered by the business. Audits also serve to demonstrate improvements made on vulnerable exposures of the organization over successive audits.

10.4 Purpose of Inspections

Inspections are often perceived to be as routine as and less important than audits. Inspections provide a great opportunity to proactively identify opportunities where trends may indicate a potential management system deficiency. For example, if during successive inspections the locks on a deenergized piece of equipment appear incorrect, this may reflect a management system deficiency related to either the procedure or the training and competency of workers. Both audits and inspections are tools that organizations may employ for different reasons. The importance and functionality of an inspection depends on the competence of the person performing the inspection. Most often, inspections conducted by leadership personnel have the greatest impact, since they possess the capacity to immediately change unsafe conditions and at-risk behaviors. A real concern with inspections is that they run the risk of becoming a check-in-the-box exercise with diminished value to the process if conducted by the same person on an ongoing basis.

Inspections can be used for the purposes in the below subsections.

10.4.1 Assessment of Compliance and Conformance

Verification of compliance and conformance is usually a tactical process employed in the field to check compliance of actual practices against preset standards and codes of practice within a certain area of operation. Inspection of confined space entry equipment or gas detection equipment may form part of a regulatory compliance requirement and would be conducted accordingly. If not followed religiously, when events occur, the organization may not be prepared, and the severity of an incident can be escalated.

10.4.2 Supervisor Development

By empowering supervisors/operatives to lead or be part of inspection teams, they get to learn more about EHS and improve their operations. Used properly, inspections may be an effective tool for personnel development for hazard and risk identification and management. Engaging inexperienced supervisory personnel in a multidisciplinary inspection team can significantly improve the capabilities of future frontline supervisory personnel.

10.4.3 Prevention of Loss

Inspections can really help identify unsafe conditions or practices in the field, preventing incidents from occurring. This is particularly important in dynamic, fast-paced work environments, where changing work conditions can lead to unplanned hazards as workers become so involved in their process that they miss the bigger challenges. Inspections that cover fire extinguishers, hazardous materials, eyewash stations, ladder use, working from heights, and vehicles have been known to prevent major incidents in organizations and must become a part of the culture of any organization that pursues OEMS.

10.4.4 Project/Construction Safety

Inspections are generally an ongoing process in project environments, and can be very effective for ensuring compliance of contractors and subcontractors in project or construction environments. Inspections in building and construction sites on a daily basis are usually recommended due to the continual changes that take place, and the emerging hazards and risks that are created through these changes.

10.4.5 Developing a Proactive Culture

Inspections can really help develop a practice of awareness for unsafe conditions and acts. Institute inspections that integrate well with leading key performance indicators (KPIs). Inspections undertaken effectively proactively identify issues that may emerge so that they can be addressed in a timely manner to prevent incidents. Inspections, when demonstrated well by leadership, have a strong impact on frontline personnel, particularly when the organization takes the time and commits resources to addressing unsafe conditions and behaviors identified during the audit.

10.5 Common Types of Audits

Organizations use different types of diagnostic tools to evaluate the strengths and weaknesses of the management system in place. Audits, when used effectively, can provide a good gauge of the health of the management systems the organization may have in place. Audits, when conducted properly, highlight both the good things the organization is doing (potential best practices) and the opportunities for improvements. The following subsections list the different types of audits applied by organizations.

10.5.1 Fire and Safety

Fire and safety audits generally look at components that support the emergency management element of the management system. Such audits would consider the design, effective emergency-related aspects, fire safety adequacy of equipment and systems, management of change (MoC) processes and practices, fire protection, and prevention system compliance. Fire and safety audits should be performed by specifically trained and competent personnel.

10.5.2 Environment and Energy Management

Environment and energy management audits are focused primarily on environmental policies and procedures, waste management, aspects and impacts registers, energy consumption audits, environmental protection and pollution prevention, compliance with local rules and regulations, and international protocol compliances. They are often governed by regulatory requirements, unless, of course, the organizational requirements exceed the regulatory requirements.

10.5.3 Occupational Health and Hygiene

Occupational health and hygiene audits are specified audits, and can look at an array of requirements from occupational health medical examinations, such as preemployment and in-employment, industrial hygiene assessments and surveys (noise, volatile organic compounds [VOCs], vibration, heat, etc.), hazard identification and communication, control of substances that are hazardous to health, and so on. These audits are particularly important for employee injury and illnesses.

10.5.4 Process Safety Management (PSM) Audits

For many large oil and gas organizations and other process organizations that have embraced the intent of PSM, the PSM audit is a highly effective method for demonstrating compliance to the requirements of each element. PSM covers all areas of process operations and, when done properly, has a hugely beneficial impact on preventing major incidents in organizations and reinforcing best practices within the organization.

10.5.5 Management System Audits

This is perhaps the most important type of audits when developing and sustaining an OEMS. Management system audits address the many components of the management system, which, from an OEMS perspective, would generally integrate EHS, quality management, PSM, and security. A management system audit is extremely powerful since it addresses all elements of the management system in a single audit. Typically, management system audits are extensive and require elaborate planning and execution. The goal of a management system audit is to identify systematic gaps in the management system so that sustainable processes and strategies are developed to address these gaps.

10.5.6 Collaborative Contractor Audits

Collaborative contractor audits generate excellent improvements in the EHS performance of contractors. Lutchman et al. (2011) advised, "When collaborative relationships are established with contractors, audits are viewed as a proactive means for identifying gaps or weaknesses in the contractor safety management systems that may result in losses or significant events" (p. 199). Today, many progressive organizations require their major contractors to follow the management systems requirements of the organization, or align with the management systems requirements.

Auditing the contractor's management system in a collaborative manner helps to move the contractor's EHS and risk management performance closer to the organization's performance management expectations. Lutchman et al. (2011) provided evidence that demonstrated the EHS improvements impact of field auditing of contractors as shown in Figure 10.1. A 50% improvement in contractor injury frequency was recorded over a 2-year period of auditing and collaborative improvement actions at Suncor Energy, Canada's largest oil and gas company.

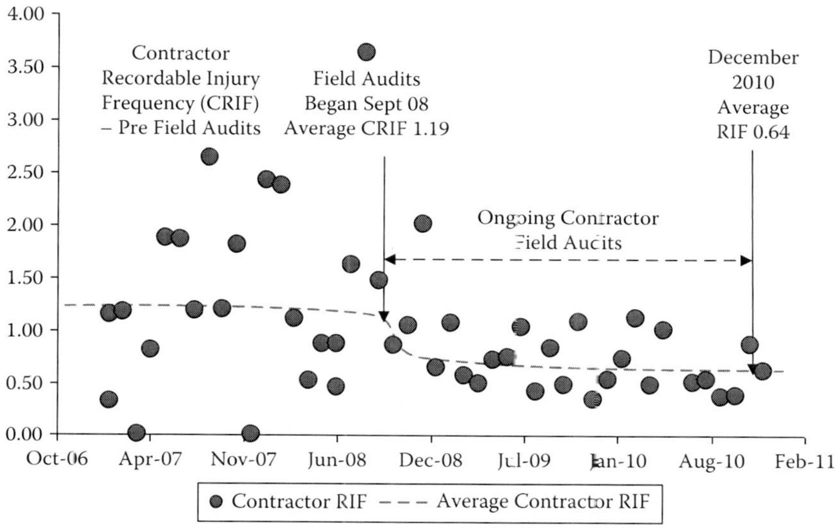

FIGURE 10.1
Impact of field audits on contractor injury frequency. (From Suncor Energy. Copyright © 2012 Suncor Energy, Inc. Approval from Suncor is required to reproduce this work.)

10.6 Challenges Faced When Conducting Audits

There are a multitude of challenges that auditors encounter during the execution of an audit. As pointed out earlier, auditors are regarded as policing agents by many business unit staff who generally turn up when something is wrong. As a consequence, therefore, auditors are required to be more open, transparent, objective, and understanding of the business unit and field operational challenges. At the onset, auditors must possess great emotional intelligence, as well as technical expertise, in order to make meaningful contributions to business performances from audits. Among the key attributes of an auditor are the following:

- Many seasoned auditors would recommend that a good auditor is a great listener. Auditors should possess great listening skills and should advise objectively, rather than use the audit as a forum for teaching, correcting, or arguing.
- Auditors are required to be well versed with the operations in general, and focus on the systems and processes with an ability to probe in a noninvasive manner.
- Auditors should be empathetic and able to evaluate complex materials and information to identify risk exposures from the information provided and assimilated.

Beyond the challenges faced by auditors are those presented by audits in general. These are listed in the following subsections.

10.6.1 Time for Auditors to Prepare Report

Generally, auditors are required to perform a balancing act between the time spent auditing and that spent in preparing the audit report. Since auditors generally function in other roles (SMEs), upon completion of the audit, it is often very difficult to have them allocate time to participate in writing and finalizing the audit report. When an audit is planned, sufficient time should be set aside for completing and finalizing the audit report before auditors return to other roles. As a rule of thumb, report generation with peer reviews, legal input, and agreement with business unit leaders on audit findings will typically require 25 to 35% of the total audit time.

10.6.2 Quality of Audit Reports

In most instances, audit reports become documents that can be used by the legal system in the event of incidents. The Texas City BP incident review and investigation pointed to a multiplicity of audit and audit reports conducted by BP. Baker et al. (2007), who produced the report on this investigation, suggested:

> The principal focus of the audits was on compliance and verifying that required management systems were in place to satisfy legal requirements. It does not appear, however, that BP used the audits to ensure that the management systems were delivering the desired safety performance or to assess a site's performance against industry best practices. (p. xv)

The report also suggested that

> BP had sometimes failed to address promptly and track to completion of process safety deficiencies identified during hazard assessments, audits, inspections, and incident investigations. The Panel's review, for example, found repeat audit findings at BP's U.S. refineries, suggesting that true root causes were not being identified and corrected. (p. xv)

Clearly, therefore, audit reports must be properly prepared, acted upon, and followed up on to ensure that value is generated from audits. A well-prepared audit report that addresses findings, complete with an adequate corrective action plan (CAP), is essential to the business.

10.6.3 Cost, Logistical, and Workload Challenges

Depending on the size and scale of operation, an intense amount of planning and logistic challenges are presented when planning and executing audits.

Travel and accommodation costs, remote locations, and intense workloads are all difficulties faced by organizations in planning and executing audits. Geographic and time zone differences place additional pressures on audit planners. Many global organizations provide auditors with additional time for travel and rest so that they can get to sites fresh and have meaningful contact time with the audited entity representatives and interviewees. This is good practice when planning audits.

10.6.4 Quality Assurance and Consistency in Report Writing

Corporations use different auditors, although many different people will have different writing styles. Quality assurance and consistency in audit report writing are essential requirements of the auditing process, so that findings are consistent among auditors. Some organizations have evolved to the extent that auditors are presented with predetermined choices and selection criteria for writing audit findings. Corrective actions are also very generic in these types of organizations, leaving the corrective actions to be determined by the business unit. Auditors should therefore be prepared beforehand to interpret and write findings in a consistent manner so that ambiguity and variations in interpretations of findings and corrective actions are removed.

10.6.5 Self-Assessment and Preparedness of the Business Unit before the Audit

Self-assessment is a recommended best practice for organizations in the management of their assets. Doing a self-assessment before an audit is a good practice. A self-assessment allows the business unit to be better prepared for the audit and to be better prepared to provide supporting information, demonstrating compliance and conformance to requirements. When a business unit is unprepared for an audit, the extent of resistance and unwillingness to cooperate with the audit team can be demoralizing to both the auditor and the business unit personnel, leading to strained relationships and mistrust. A well-planned audit with the key objectives and protocols communicated with adequate lead time is essential. A schedule for the auditors with sufficient time to rest, work, and write their reports contributes tremendously to successful and meaningful results.

10.6.6 Audit Fatigue

Audit fatigue is a challenge faced by many businesses. Striking the right balance between the numbers of audits conducted and the value derived from such audits is a challenge faced by many organizations today. How often is too often for executing audits? Adequate time must be allowed between audits for the organization to activate its CAP and steward the CAP prior to follow-up audits.

10.6.7 Timely Feedback and Corrective Action Plans

Receiving a CAP from the business to address findings is often among the most difficult aspects of the auditing process. Stewardship of gap closure activities is also a challenge for most businesses. Some of the more progressive organizations have established verifier roles within the audit organizations to ensure effective stewardship of the CAP. A key to success in developing the CAP is to ensure an effective and comprehensive audit closeout meeting in which the site leadership is intimately involved and engaged. The closeout meeting should detail follow-up actions and articulate the value generated by the audit. Critical to success, however, the business unit leaders must agree that the findings of the audit are real and present the business risk determined by the audit.

10.7 Audit Triggers

Audits are generally triggered by schedules or unacceptable business performances. For example, an EHS audit may be determined by trends in incident numbers and severity being experienced by the organization/contractor during the execution of assigned work. Unplanned audits, as opportunity permits, may be required due to operational priorities, regulatory requirements, new contracts, and outcomes of internal or third-party audits. Management system audits may be conducted to gauge progress of the organization relative to the baseline established for the desired state of the organization.

10.8 Preparing for and Conducting a Management System Audit

A management system audit is an intensive process that should be started several months ahead of the planned event. There are multiple steps in the audit to ensure success. The details of each step are provided in Table 10.1. Classifying findings from the audit and determining corrective actions are important components of the audit process. Figure 10.2 provides the procedure, accountability, competency, assurance (PACA) framework for classifying findings and suggesting corrective actions for such findings.

Findings may be classified into the following:

1. Single-point findings: Gaps or findings are easily closed; they are localized and do not impact the organization in a general way.
2. Systematic findings: Systematic findings and gaps are more typical of management system findings and gaps. When a systematic finding is found in one business unit, the same finding or gap is

TABLE 10.1

Steps in the Audit Process

Audit Step	Activity	Tools
• Preparation and planning	• Notify business unit/organization of audit and scope of the management system audit. • May cover all elements of the management system or selected high-priority elements. • Select auditors specific to the scope of work requirements. • Verify logistics—transportation, flights, accommodation, access to Internet—are looked after.	• Audit scheduling and planning tool
• Preread materials preparation	• Work with business unit/organization to accumulate required preread materials (procedures, reports, standards) that demonstrate compliance and conformance to the management system requirements. • Provide auditors access to preread information.	• Common access drive for all auditors
• Opening meeting	• 1- to 1.5-hour meeting with site leadership to reconfirm scope of the audit. • Clearly articulate proposed value of the audit to the business. • Seek specific area challenges from leadership so that the auditors can provide assistance based on findings. • Create a collaborative atmosphere. • Create an open forum for discussion among stakeholders.	• Presentation that articulates why audit is required if not requested by the business
• Data collection	• Preparation—collect documentation from business unit/organization and review against management system element requirements. Can be done in advance of the audit. • Conduct interviews. • Perform field observations.	• Follow audit protocols for management system requirements • Management system audit checklist and evidence guide

(Continued)

TABLE 10.1 (CONTINUED)

Steps in the Audit Process

Audit Step	Activity	Tools
• Analyze data	• Identify gaps in the business unit/organization management system. • Classify findings into single point vs. systematic. • Categorize and risk rank findings.	• Template for comparing evidence vs. requirements
• Write findings and corrective actions	• Write findings to document gaps identified. • Write corrective actions. • Define action requirements and due dates. • Review preliminary report with element business unit/organization on a daily basis.	• Audit report template
• Audit closeout meeting	• Thank the business and participants for their help and support. • Highlight high-priority gaps—risk exposures. • Highlight best practices and successes. • Agree on delivery of time frame of finalized audit report. • Agree on timing and delivery of the finalized and resourced CAP. • Agree on all other follow-up actions.	• Presentation that articulates best practices and opportunities for improvements—gaps in the management system
• Distribute audit report for review	• Review with all relevant stakeholders (including contractor). • Finalize with input from the organizational legal team. • Agree with stakeholders on corrective actions. • Issue finalized report.	• Action management spreadsheet/action tracker
• Action management and gap closure stewardship	• Review proposed gap closure strategy and action plan. • Agree on schedule for follow-up reviews and actions. • Consider the use of verifiers to ensure gaps are closed in the agreed time frames.	• Develop escalation and agreed consequences action plan

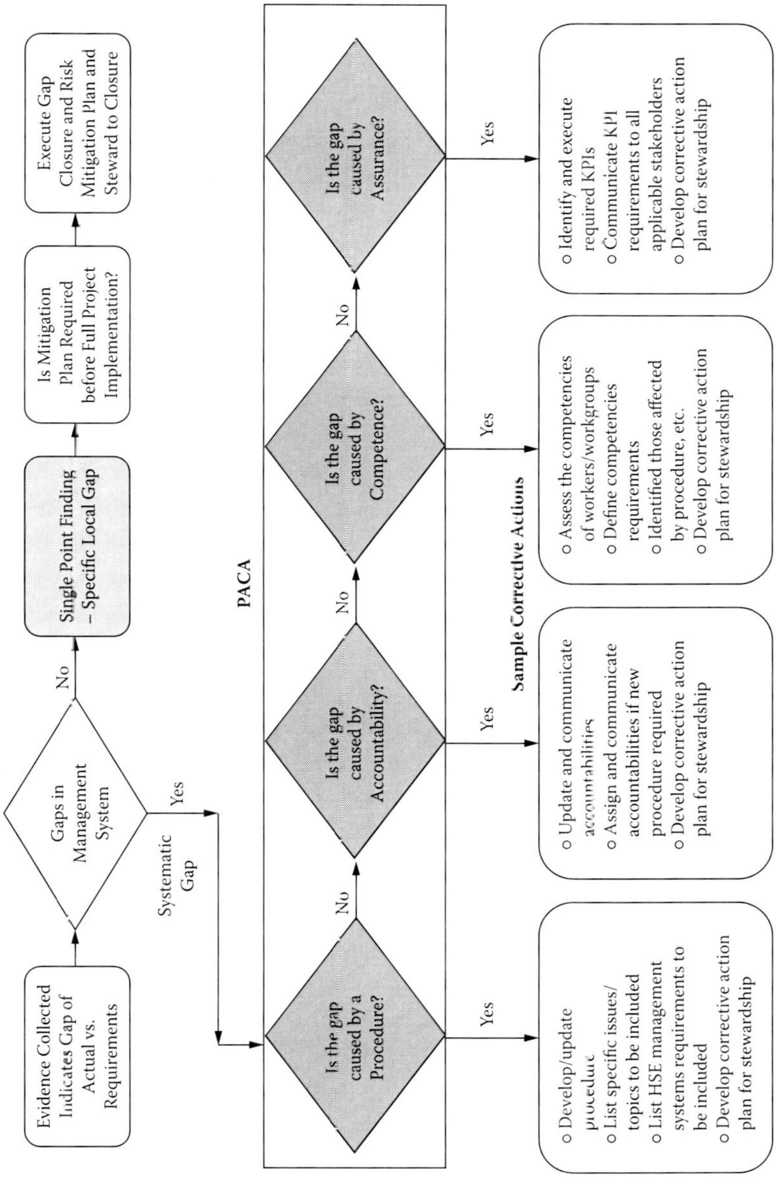

FIGURE 10.2
Classifying findings and proposed actions.

likely to be found across multiple business units. Such gaps require a complete evaluation of the process and the application of a series of actions to ensure the gaps identified are closed. For example, a PACA finding may require the need for a new or updated procedure that requires a MoC process be followed. This may further entail training users, follow-up training, assurance verification, and assigning accountabilities for the entire process.

Care must be taken to ensure that all findings are assigned the right corrective action. Failure to do so will result in a perpetuation of the systematic gap throughout the management system that will appear once again in subsequent audits and will have failed to address the risk exposures of the organization. Often, an auditor peer review of both findings and corrective actions will help refine and upgrade the quality of the finding and corrective actions. Care must also be taken by the auditor to avoid lumping findings together that may cause confusion to the business unit in crafting the CAP. Findings should be specific, and corrective actions should be specific, measurable, achievable, realistic, and time bounded (SMART) and verifiable.

10.9 Conclusions

Audits and inspections are very important proactive and leading EHS processes for any organization—especially for high-reliability/high-risk industries on the quest for operations discipline and OEMS. Practitioners are encouraged to see auditing as a proactive measure to prevent incidents before they occur and as a means of improving business performance proactively. Auditors who understand the value of collaborative auditing should be developed within the organization. They must be trained in creating win-win outcomes, whereby the audited organization does not feel violated or unfairly treated in the process. Collaborative auditing has the potential to transition organizations toward operation discipline and excellence while developing a proactive safety culture.

The development and need for a specialist team at the management system audit function can be extremely beneficial to the organization. Credibility derived from competence, experience, fair and transparent auditing practices and behaviors, empathy, collaboration, and communication are desired behaviors and are characteristic of the auditor. Well-developed and -tested audit protocols help guide both the business and the auditor in the auditing process.

Once embraced as an improvement process by the business unit and organization, audits provide a powerful path forward for improvements in organizational and contractor's EHS performance When used correctly, audits and inspections can be one of the most powerful tools in developing the organization's safety culture, and transition it from a good safety culture to an operationally excellent safety culture.

References

Baker, A.J., III, Levenson, N.L., Priest, S., Tebo, P.V., Rosenthal, I., Bowman, F.L., Hendershot, D., Wilson, D., Gorton, S., and Weigman, D.A. (2007). The BP U.S. Refineries Independent Safety Review Panel. Retrieved April 4, 2014, from http://www.bp.com/liveassets/bp_internet/globalbp/globalbp_uk_english/SP/STAGING/local_assets/assets/pdfs/Baker_panel_report.pdf.

Lutchman, C., Evans, D., Maharaj, R., and Sharma, R. (2013). *Process safety management: Leveraging networks and communities of practice for continuous improvement*. Taylor & Francis, CRC Press, Boca Raton, FL.

11

Fundamental 6: Closing the Gaps—Discipline

Once the baseline assessment has been completed and identified gaps closed, the organization will necessarily transition into a mode of sustainment or continuous improvement depending on the vision of the organization. Most organizations embark on a process of continually pushing themselves for higher and higher performance. Therefore, as performance targets and element requirements change over time, it's necessary to assess the performance of the business on an ongoing basis.

As discussed in fundamental 5, management system audits are performed to determine the levels of conformance and compliance by the business to internal and regulatory requirements of the organization. Management system audits seek to identify systematic gaps in the management system that, if not addressed, perpetuate over time, exposing the organization to significant risks and potential for experiencing a major incident.

Upon completion of a management system audit, the organization should develop a gap closure plan for addressing all systematic gaps identified. In this section, the authors recommend the need for operational discipline, whereby standardized procedures and practices are used to address similar gaps across multiple business units. Operations discipline is critical at this stage since leaders tend to operate on the principle that we have achieved organizational targets and view incremental improvements as requiring too much effort to do so.

11.1 Addressing Gaps Identified

As discussed in fundamental 5, audit findings may fall into single-point findings (gaps) or systematic findings (gaps), both of which require distinctly different approaches and levels of discipline for resolution.

11.1.1 Single-Point Findings and Gaps

Such gaps require simple solutions, which, once executed, may eliminate the risk of exposure completely. This type of finding may require simple solutions, such as the following:

1. Installing a mechanical guard on rotating equipment
2. Installing permanent access to a vessel or piece of equipment
3. Providing engineering solutions to a particular risk exposure
4. Gap closure solutions that are normally not resource-intensive in closing them

Single-point findings and gaps are generally the outcome of a small human error—oversight that leads to a gap and a specific risk exposure.

Several single-point findings or gaps may exist, and consistent with good business practice, prioritization is required to manage resources demands and workloads. The simplest form of prioritization is in the risk ranking of risk exposures resulting from the gap. When risk ranked, high risk exposures should be addressed first, followed by medium or moderate risk exposures, and finally low risk exposures.

11.1.2 Systematic Findings and Gaps

Systematic findings and gaps are indicative of a more deeply rooted problem or risk exposure to which the organization or business unit is exposed. Systematic findings and gaps require the following for addressing them:

1. A structured and strategic planned approach for closing the gap
2. Perhaps a working team to analyze the gap, understand the risk exposures, and develop the right strategies for mitigating risk exposures
3. Conducting a stakeholder impact assessment
4. Risk ranking of the identified gaps
5. Closing gaps that are immediately dangerous to life and health as a first priority
6. Applying the opportunity matrix to prioritize gap closure work
7. Identifying and exploiting low-hanging fruits, for example, gap closure opportunities that are characterized as follows:
    - Subelements requiring small or incremental efforts to achieve compliance or conformance
    - Situations where a work practice may exist, but processes are not formalized
    - Documentation may be required as a means for achieving compliance

Systematic gap closure requires the application of a plan–do–check–act (PDCA) model to ensure work is properly planned, executed, verified, and corrective actions taken in a timely manner.

11.2 Operational Discipline and Leadership Commitment

Prioritizing and closing systematic gaps requires a tremendous amount of operations discipline by the organization. Why is operations discipline essential? Operations discipline is required because of the many competing business priorities of the organization or business unit. Leaders are challenged to balance budgets, generate profits, cut cost, increase reliability, inspire and motivate the workforce, and become leaner, while at the same time attempting to prioritize and close gaps in the management system. The irony of it all is that these gaps and risk exposures existed while the organization continued to operate without incident and without leaders being aware them. Operations discipline by leaders and the workforce is essential in order to be motivated and inspired to want to close the gaps and risk exposures. Closing these gaps requires leadership commitment and support.

11.2.1 Leadership Commitment

Leadership commitment in any business process is essential for success. Where closing the gaps is concerned, leadership commitment is demonstrated in the following ways:

- Leaders are effective in removing barriers so that gaps can be effectively closed.
- Leaders provide support in terms of guidance and motivation to ensure the workforce continues to be focused on the strategic goals.
- Adequate resources are allocated to avoid frustrating stakeholders involved in gap closure work.
- Stewardship and accountability while serving as the principal interface with senior leadership on an ongoing basis.

Commitment is unquestionable when a charter is established for the working team, and the charter is sponsored by a senior leader of the organization or business unit.

11.2.2 Team Charter

A team charter seeks to hold the team accountable for the scope of work defined in the charter. A sample one-page charter is provided in Table 11.1.

A charter is created for each gap or group of gaps that the team may be required to close. Systematic gaps may be grouped together for corrective actions if there are synergies to be derived in solving them together. For example (referring back to Figure 8.1), gaps were found in the following

TABLE 11.1

Sample One-Page Charter for Gap Closure

Measurable Goals and Objectives	Critical Success Factors	In Scope	Out of Scope
Deliverables	**Meeting Frequency and Norms**	**Team Members and Tenure**	**Leadership Sponsor and Approval**

Source: Lutchman, C. et al., *Process Safety Management: Leveraging Networks and Communities of Practice for Continuous Improvement*, Taylor & Francis, CRC Press, Boca Raton, FL, 2013.

subelements of the element event (incident) management and learning as follows during the baseline self-assessment (BSA):

- Incidents and near misses are viewed as learning opportunities for the organization and industry as a whole.
- Effective root cause analysis (RCA) is conducted on all incidents. Corrective actions are generated and stewarded to closure.
- The RCA process seeks to avoid blame assignment and finger pointing, while focusing on identifying management system deficiencies, to prevent recurrence of incidents.
- The extent of the RCA conducted is dependent upon the severity of the incident or the potential severity from near misses.
- Knowledge generated from incidents is shared within the organization. Where applicable, this knowledge is shared in an organized manner within the industry and across industries to prevent repetition of the same or similar incidents.

Had these or similar gaps been found during the audit, and had they been present across the organization or several business units, a team chartered with responsibilities to develop a strategic plan and execute the same for addressing these gaps collectively may have been required. Table 11.2 provides a sample of deficiencies associated with each subelement's requirements and potential solutions (which become the deliverables in the charter).

TABLE 11.2

Subelements, Gaps, and Potential Solutions (Deliverables in the Charter)

Subelements	Audit Findings—Gaps	Potential Solutions (Deliverables in Charter)—Remember PACA

• Incidents and near misses are viewed as learning opportunities for the organization and industry as a whole.	• Not all incidents and near misses are being thoroughly investigated.
	• The organization does not have time to learn from industry incidents.
		• Develop a procedure for incident management that addresses the following:
		  • Accountabilities, competencies, and assurance requirements.
		  • What types of incidents shall be investigated? Link to the organization's risk management process and risk matrix.
		  • Corrective action management process that includes accountabilities and follow-up actions in the event of delays.
		  • Who should be involved in incident investigation—minimum team composition?
		• Select a single/preferred RCA tool/process to be used by the organization. RCA tool/process should identify root causes as to the management system failures level.
		• Develop a standardized template for sharing knowledge in environmental health and safety (EHS).

• Effective root cause analyses (RCAs) are conducted on all incidents. Corrective actions are generated and stewarded to closure.	• The organization uses multiple RCA methodologies. • There is no consistency in the findings generated from similar incidents. • There is no documented follow-up process for corrective actions management.	
• The RCA process seeks to avoid blame assignment and finger pointing while focusing on identifying management system deficiencies to prevent recurrence of incidents.	• People who perform RCA are not always trained to do so. • RCA generally falls short of determining root causes and management system deficiencies.	
• The extent of the RCA conducted is dependent upon the severity of the incident or the potential severity from near misses.	• There is no consistent process for determining incidents that are investigated vs. those that are not.	
• Knowledge generated from incidents is shared within the organization. Where applicable, this knowledge is shared in an organized manner within the industry and across industries to prevent repetition of the same or similar incidents.	• There is no organized process for sharing learnings within the organization. • There is no organized process for sharing within the industry nor is there learning from the industry—learning is ad hoc.	

(Continued)

TABLE 11.2 (CONTINUED)

Subelements, Gaps, and Potential Solutions (Deliverables in the Charter)

Subelements	Audit Findings—Gaps	Potential Solutions (Deliverables in Charter)—Remember PACA
	• There are no standardized processes for sharing learning, nor are there processes preventing the transfer of sensitive information outside the organization. • Learnings are shared only within the health, safety, and environment (HSE) organization with little transfer to other stakeholders within the organization.	• Develop a process for sharing learning in EHS within the business unit, the organization, the industry, and across industries. • Develop a training package and conduct training across the organization for incident management (IM) and the procedure. • Execute training across the organization—awareness and competency training for appropriate stakeholder groups.

11.3 Execution of Work to Close Gaps

As with all work, effective planning leads to smoother and more efficient execution of any work. Execution of the work, much like that described in fundamental 3, requires the organization to follow the principles of Demming's PDCA model. Execution of this work is intended to drive sustained improvements in the business processes, and therefore requires careful attention to detail and must be regarded as a required process of the business.

The key thing to remember is that audits identify gaps in the management systems that can potentially lead to incidents or suboptimal performance. Furthermore, when audits identify findings, the following chronology of events is required:

1. Management system gaps are generated or identified.
2. The auditor or audit team makes recommendations to the business unit or organizational leadership for closing the gap.
3. The business unit or organizational leadership:
   a. Identify the potential solutions for closing the gap
   b. Evaluate the potential solutions for closing the gap and select the most approptiate option from them
4. Development of a plan/strategy for closing the gap:
   a. Execution of the plan/strategy for closing the gap
5. Follow the PDCA model to determine the effectiveness of the gap closure plan or strategy.

When developing the plan or strategy for closing the gap, a key requirement for long-term success is stakeholder involvement, engagement, and communication that create the required ownership for sustainment. The simplified PDCA model discussed in fundamental 3 provides the required framework for sustainment of the gap closure plan or strategy. As with all plans, there must be leadership, resources, leadership commitment, key milestones, and periodic assessment for course correction as required.

11.4 Conclusion

This fundamental is among one of the more important fundamentals since it requires the operations discipline essential to close identified gaps. Leaders can become complacent in their commitments, particularly when there are multiple priorities faced by the organization. Care must be taken, however, to remind leaders that stakeholders are quite forgiving when the business is ignorant or unaware of potential gaps in the management system.

However, when gaps have been identified and leaders fail or procrastinate in closing them, and incidents occur from failing to respond to them, the consequences can be quite severe from a stakeholder perspective. Therefore, as a reminder, business units or organizational leadership should consider audit reports as legal documents and respond proactively to take corrective actions and steward these corrective actions to closure. This type of behavior requires strong operational discipline and leadership to ensure the sixth fundamental of an operational excellence management system (OEMS) is met.

12

Fundamental 7: Networks and Communities of Practice for Continuous Improvements and Shared Learning

12.1 Introduction

Studies show that 90 to 95% of workplace incidents are avoidable and 80 to 85% are repeated. Learning from events is an effective means for reducing repeat incidents and cost while improving operating reliability and performance. Learning from incidents, knowledge generated from best practices, and continuous improvements seldom make it to the front line where such learning can be beneficial in preventing incidents. More importantly, shared learning is the last remaining low-hanging fruit for sustained improvements in reliability, reduced incidents in the workplace, and improved organizational performance. In this chapter, the authors explore and discuss the following:

- Organizational challenges experienced in generating and sharing knowledge
- The maturity journey in setting up networks and communities of practice (CoPs)
- Leading networks and CoPs and the governance of multiple networks and CoPs
- Value maximization opportunities and expected performance results from networks and CoPs

The authors argue that in order to move learnings from lessons to applied learning, the following key requirements are essential:

1. Learning must be structured and presented to cater to both Generation X and Generation Y. Both groups learn differently, and their learning needs must be met for success.
2. Approaches to generating and sharing learning must be simple and organized for sustained improvements.

3. Learning may come from several sources, including internal learning, learning from peers (within industry), and learning from other industries, as well as from sources such as research and investigations done by specialized organizations, like the Chemical Safety Board (CSB).
4. Learning activity must go beyond safety moments and emails. Learnings must be embedded in knowledge repositories, procedures, and standards. In addition, practitioners must be aware and trained.

12.1.1 Current State of Generating and Sharing Knowledge

In many organizations, most of the lessons learned from past and current events occur after the events and are considered a reactionary or lagging indicator focus of safety performance. For many organizations (even in large, deep-pocket organizations), generating and sharing knowledge from incidents continues to be weak and disorganized. Often, we fail to develop internal systems and processes that can lead to the development and full sharing of knowledge generated from within the organization. A summary of the market where shared learning is concerned is demonstrated in Table 12.1 (Lutchman, 2012).

In many instances, leaders tend to adopt protectionist approaches to data, information, and knowledge among similar business units in an

TABLE 12.1

Market Summary of Current Knowledge Sharing

Sharing of Knowledge	Market Situation
Within organization—internal (business units and functional areas)	**Early stages of development:** Organizations are in the early stages of assessing the value of shared learning and are now evaluating how to share knowledge more effectively. There are no standardized processes for sharing of knowledge. Emphasis continues to be on sharing of data and information, as opposed to knowledge that is beneficial to organizations.
Within industry	**Disorganized and indiscriminate:** Sharing within industries is disorganized and indiscriminate. Information is shared informally among peers via email in a disorganized manner with no concerns of the type and quality of information being shared. There is very little regard for sensitivity of the information being shared, as well as the accompanying liability in sharing.
Across industries	**Not available:** Very little, if any, sharing of knowledge occurs across industries.

Source: Lutchman, C., How to Go from Lesson to Learned: PSM from Engineering to Operations, presented at the 8th Global Congress on Process Safety, Houston, TX, April 1–4, 2012.

organization. As a result, communication and interactions between similar business units (business areas, operating areas) of an organization may be limited, and are often nonexistent within the same organization. While this concern is challenging within the organization, it is even more difficult to address between organizations in the same industry because of weak collaboration and competition regulations.

12.2 Are We Unable to Learn?

Kletz (2001) suggested that understanding how incidents occur requires incident investigators to continue to ask the question *why* while undergoing an investigation process much like peeling the layers of skin of an onion. From his early work in investigating incidents, Kletz (2001) pointed out the common underlying causes of incidents, to include the following:

- Protective systems failure
- Poor procedure and poor management

Today, these underlying causes of incidents continue to be very prevalent among the leading causes of incidents.

Hopkins (2012), in a review of the BP Gulf of Mexico Macondo incident, pointed to protective system failures, as well as organizational and managerial issues. According to Raleigh (2013), the goal of continuous improvements to avoid repeating the Macondo and similar incidents reflects fundamental challenges faced by organizations to implement procedures designed to address the myriad often complex elements and interactions characteristic of process safety management. Raleigh (2013) also suggested that organizations have short memories. They respond well to safety issues when major incidents occur, and revert to other business priorities once easy-to-implement improvement measures have been activated.

Table 12.2 highlights some repeat major incidents that demonstrate the challenge and apparent reluctance of organizations to learn from peers and other industries in the repeat of similar combustible dust incidents. This pattern of repeat major incidents is demonstrated in several areas of process safety management that are highlighted by the U.S. Chemical Safety Board (2014), such as

- Hot work permits—procedures
- Gas blows—procedures
- Runaway chemical reactions—process operations
- Fires and explosions—mechanical integrity and maintenance management procedures

TABLE 12.2

Repeat Combustible Dust Major Incidents

Major Combustible Dust Explosions	Date	Impact
Sugar dust explosion and fire: Georgia	February 2008	14 killed, 36 injured
Metal dust fire and explosion: Indiana	October 2003	1 killed, 1 injured
Organic dust fire and explosion: Kentucky	February 2003	7 killed, 37 injured
Organic dust fire and explosion: North Carolina	January 2003	6 killed, 38 injured
Organic dust fire and explosion: Massachusetts	February 1999	3 killed, 9 injured
A series of devastating grain dust explosions in grain elevators	1970s	59 killed, 49 injured

Source: Safety Erudite, CCS: Collect, Standardize, and Share, Our High Value Low Effort Transformational Approach to Improving Safety in Your Workplace through Shared Knowledge, 2014, retrieved December 27, 2014, from https://www.safetyerudite.com/.

As organizations continue to repeat the same and similar incidents on an ongoing basis, the question regarding our ability to learn arises. What prevents us from learning to avoid repeating the same and similar incidents? Do we fail to learn because

- Organizations are indifferent to the incidents experienced by other organizations? We adopt the "it can only happen to others; we are better than that" mentality.
- Organizations do not know how to learn proactively from within?
- The cost of internal learning is too high, so owners avoid required investments for doing so?
- The leadership challenge to internal learning—organizations have become so lean that leaders have no time to learn from internal sources?

Regardless of the reasons for weak learning from internal and external sources, the apparent message to stakeholders is that many organizations are immune to learning. This view (despite tremendous improvements in personnel safety) is very apparent from the repeat of the many major process safety incidents highlighted by the U.S. CSB and demonstrated in Table 12.2.

12.3 Network Structure

What are networks? Historically, we have used networks to share information and knowledge, enable changes in organizations, and support our personal and social interests. In this chapter, the authors define a process safety management (PSM) network as a group of experts (subject matter experts (SMEs)) who are brought together as a team with a fixed mandate of facilitating improvements in an assigned area of focus for the business.

Networks may be formal or informal; however, the modes of operation continue to be the same. From a PSM perspective, several networks are required to generate continuous improvements from learnings. A key requirement is the need for integration among networks such that there is continuous cross-pollination across networks to avoid duplication of efforts and a consequential wastage of resources resulting in potential conflicts. Typically, networks are comprised of a core team, supported by a wider group of SMEs that interfaces further still with a community of practice team

12.3.1 Core Team

The core team is responsible for converting data and information into learnings and knowledge, and aligning network focus to support business priorities. This team is generally comprised of senior specialists and leaders of the organization who possess the ability to analyze alternatives for creating value, maximizing best practices, and learning. This team is continually engaging with the organizational SMEs, seeking ways and means to improve and address persistent and new problems, as well as identifying best practices that can be shared across the entire organization. The core team must be empowered by the organization to challenge the status quo and generate change in the organization. Typically the core team is chartered and comprised of four to eight employees who are motivated to address the tasks at hand. An important success factor is that core team members want to be a part of the network and are passionate to make things better. Most importantly, the network members need to be supported by their supervisors to commit time to network activity.

12.3.2 Subject Matter Experts

SMEs are a pool of resources across the enterprise that possess expert knowledge in the area of focus of the network. SMEs are a key resource pool for the core team in providing expert knowledge to the network. They also form the conduit for activating and supporting new learning and best practices as they are rolled out across the enterprise. The SME pool may vary in size depending on the scale of operation of the enterprise or the amount of business units/areas the organization may have. Balanced business unit/area representation and voice helps in improving contributions and executing new knowledge and learnings.

12.3.3 Community of Practice Team

The community of practice team is comprised of interested stakeholders across the enterprise who are involved in the area of focus of the network.

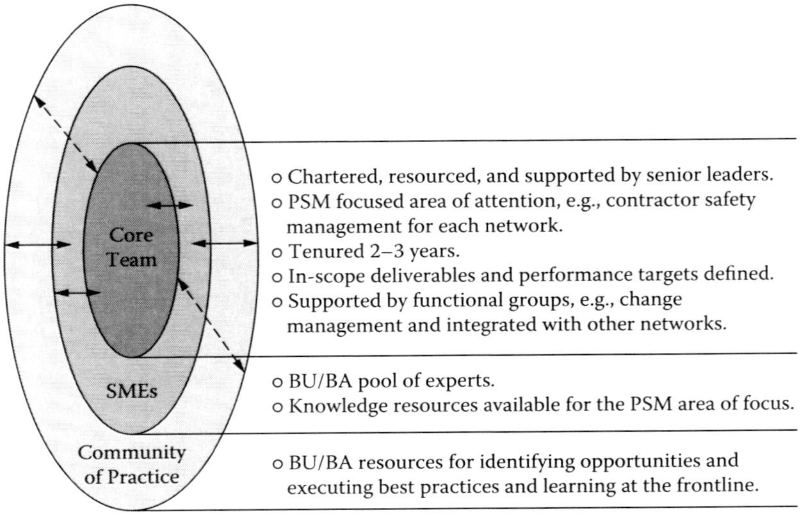

FIGURE 12.1
Sample Network Structure. (From Lutchman, C. et al., *Process Safety Management: Leveraging Networks and Communities of Practice for Continuous Improvement*, Taylor & Francis, CRC Press, Boca Raton, FL, 2013.)

Members of this team are the doers, and are responsible for executing best practices, new knowledge, and learnings across the enterprise. The size of the community may vary based on the scale of operation and the priority of the focus area.

Figure 12.1 provides a sample network structure with key stakeholder groups for generating continuous improvements and learnings.

12.4 Organizational Challenges to Generating PSM Knowledge from Networks

Focused attention to PSM has brought about tremendous improvements in process operations and organizational performance. However, PSM in many developing countries continues to be an elusive vision. In Canada, PSM is proactively being adopted by organizations to improve reliability and enhance existing management systems. As an extension of PSM, implementation networks are being created to generate and transfer solutions to simplifying complex process and eliminating managerial problems into applied knowledge at the front line.

Lutchman et al. (2013) highlighted challenges to generating new knowledge and transferring this new knowledge to the front line. They also highlighted challenges faced by organizations in managing networks. In this chapter, however, the authors point to additional business challenges faced by organizations that are generally overlooked when setting up networks. These challenges, unless properly addressed, hinder the formation of networks, and defer accompanying benefits. Among these challenges are the following:

1. Leadership resistance
2. Silo organizations and organizational skeptics
3. Initial investments and activation expenditures
4. Availability of skilled, trained, and competent network members
5. Formal network structures vs. informal CoPs
6. Time commitments from SMEs

12.4.1 Leadership Resistance

Perhaps the primary challenge faced by organizations in setting up networks is leadership resistance. In the oil and gas industry in particular, many leaders believe that refining technology has been perfected over time and there is very little, if any, opportunity for improvements. Against this background, it is easy to understand why leaders resist the formation of networks.

It is noteworthy, however, that leadership resistance tends to be experienced during the initial setup and activation of networks. Leadership resistance is reduced and often removed when the network steering team undertakes extensive communication and sharing of the value proposition of networks that leads to the creation and sharing of a common vision.

12.4.2 Silo Organizations and Organizational Skeptics

Among the leading challenges faced by organizations in setting up networks are what the authors' term silo organizations and organizational skeptics. In many large, geographically distributed and diversified business structures, the establishment of silo organizations hinders the formation of networks. In these silo organizations, leaders prefer to operate independently of the rest of the organization and do not wish to be disturbed by corporate interference or other organizational distractions.

Within these silo organizations, leaders are comfortable with the way they operate and do not see the benefit of changing. Such silo organizations are generally characterized by few problems, and leaders do not have a reason to change. More often, silo organizations advocate that other organizations adopt their operating models while disregarding the benefits of synergies, collaboration, and continuous improvements generated from learning from each other.

Silo organizations are transformed into strong supporters when leaders recognize the collective benefits for the organizations as a whole. This requires extensive communication by the network steering team and the sharing of the organizational vision.

12.4.3 Initial Investments and Activation Expenditures

Networks can be quite costly, and there is also a lag period between network setup and value creation. With all of the preparatory work required for communication, member selection, logistics management, and travel to various business sites, the initial and sustaining investments for setting up networks can be seen as nice to have and a drain on limited operating expense budgets.

Cost is multiplied further when there are multiple networks to be established. Indeed, when one considers the range of networks that are potentially required in an operating facility, it is easy to see how initial investment costs can become a major barrier. Typical networks found within an organization may include the following:

1. Rotating equipment
2. Process hazard analysis
3. Mechanical integrity
4. Process safety information
5. Electrical management
6. Pressure vessels
7. Prestart-up safety review
8. Management of change
9. Contractor management
10. Incident management
11. Emergency management
12. Process technologies, e.g., coking, hydrotreaters, etc.

This list is not exhaustive, and networks may be created based on organizational needs for generating solutions to immediate problems and for taking advantage of longer-term business opportunities. Ultimately, the work of networks is manifested in operating performance improvements and continuous business improvements.

12.4.4 Availability of Skilled, Trained, and Competent Network Members

A common factor that impedes the development of most networks is the availability of skilled, trained, and competent personnel required to staff the network. Lutchman et al. (2013) suggested the need for subject matter experts (SMEs) for staffing the core teams of the network. As organizations have grown leaner over time, most personnel are now stretched to the limit, and taking on additional responsibilities can be a huge challenge to the worker as well as to organizational priorities. Furthermore, organizations are often challenged to release their experienced and qualified personnel and would often seek to allocate less skilled, experienced, and competent resources to networks while they retain higher-quality resources for assurance purposes.

It is important to note that the best knowledge generated from networks is generated when SMEs are able to challenge each other to produce higher-quality solutions. While participation in a network can be a learning opportunity for less qualified personnel, this should not be the norm. The focus and strength of the network is diluted when competencies are diluted.

The availability of skilled and competent specialists is further constrained by the time commitments required for contributing to the network's deliverables. The greater the time commitments required by the network (which can vary from 10 to 25% of annual work time), the more difficult it is to staff the network.

12.4.5 Formal Network Structures vs. Informal Networks and Communities of Practice (CoPs)

Networks can function as both formal and informal workplace entities. A formal structure with governance and stewardship by a steering committee generates more consistent and sustainable results. The steering committee provides senior leadership support and visibility, as well as network integration and alignment to business priorities. Informal networks and communities of practice tend to start strong with passion and interest level among the members, but become rudderless and tend to drift into disuse or nonparticipation because organizational priorities and people's interests change. In the absence of formal network structures, a well-coordinated approach for generating knowledge and improvements is missed.

12.5 Networks Maturity Journey

Networks are generally long-term operating teams within the organization. Once networks are established and activated, the time leading to maturity may vary based on the priorities of deliverables. In situations where the organization may have identified existing bad actors (recurrent problems of the same types and origins) that are to be addressed as priority deliverables, the network maturity period may be quite short. Support functions are also critical in the maturity journey for networks.

Support groups, like change management, corporate communications, and knowledge management, can assist the networks greatly in improving their effectiveness. Network effectiveness is influenced greatly by their ability to influence change, communicate successes and learnings, as well as capture and transfer knowledge efficiently across the organization. In this section, the authors share and address the steps involved in setting up and activating the network through maturity.

12.5.1 Selecting Network Members—Core Team, SMEs, and Communities of Practice

Since all members of the network originate from different parts of the organization, one would think it would be fairly easy to pull together all team members of the core team, SMEs, and the CoPs. The reality, however, is that this process can be extremely lengthy since a formal process must exist in achieving the following:

- Establishing the right levels of expertise and experience of the various team members within each level (core team, SMEs, and the CoPs) of the network
- Receiving approvals from member's leaders for their participation in the network
- Since network membership requires people who want to be there, there is an initial lag time in getting participation
- Selecting core team members with similar experience and workplace performance.
- Selecting core team members who possess nonauthoritative leadership and creative thinking capabilities

Once the selection of network members (core team, SMEs, and CoPs) is complete, the next step in the maturity journey involves a chartering process.

12.5.2 Chartering the Network

A formal, simple charter commits the core team to its deliverables and enables the team to be focused on specific, prioritized deliverables. Charters enable the core teams to develop specific work plans complete with timelines and key performance indicators (KPIs) for each measurable deliverable. This charter also holds the core team accountable.

A simple, one-page charter is best. It serves as a quick reminder to the team of its network priorities. More importantly, from a network governance perspective (senior leadership team that enables all of the networks set up by the organization), simple charters help in the stewardship of multiple networks. Table 12.3 provides a simple one-page charter for networks.

12.5.3 Network Kickoff and Goal Setting

When activating networks, a well-organized PSM network kickoff conference helps in the simultaneous kickoff of all networks. A kickoff conference helps in setting the tone for all networks. It also provides the right forum for communicating to all core team members the critical information regarding goals, priorities, leadership commitments, and fundamental information indicating how these networks will function going forward.

A kickoff conference also provides opportunities for members of different networks to get to know each other and recognize network overlap and collaboration opportunities such that the duplication of effort is avoided, and opportunities for synergies are leveraged. The most valuable benefit of the kickoff conference is that all members hear the same thing at the same time and communication confusion is avoided.

12.5.4 Establishing Work Plans and KPIs

Once a network has been activated, the next step is to develop a work plan to meet the deliverables defined in the charter. All members of the core team are involved in this process. When developing the work plan, the core team may include the following:

- Resources required
- Meeting frequencies and other network collaboration requirements
- SMEs to be consulted
- Current status assessment and gap analysis to achieve desired state
- Communication strategies
- Strategies for rolling out solutions to the organization
- Timelines for delivery of solutions
- Subcommittees required and distribution of work among core team members

A well-documented work plan helps to keep the network on track.

12.5.5 Network Leaders Integration and Forum

As part of the stewardship process, a valuable component of the network maturity process is the need for bringing together all network leaders in a

TABLE 12.3

Simple One-Page MoC Network Charter

Purpose: Network the (company name)-wide community that deals with management of change (MoC) (excluding MoC-people) to drive operational excellence and to do so while leveraging the knowledge and expertise that resides within the organization and throughout the industry.
Time frame: Steering team to review network effectiveness monthly.
Commencing: (Insert kickoff date here.)

Objectives	Measurable Goals	Key Activities	Tracking Measures (How)	Critical Success Factors
• Support/drive operational excellence in management of change (MoC). • Compare, contrast, and reconcile best internal operating practices between business unit (BU) and business area (BA)/across the network. • Identify areas of opportunity for continuous improvement of the MoC standards. • Support implementation of the standards across the network. • Provide a forum to facilitate these objectives efficiently and effectively. • Provide guidance to the network on the interpretation of standard.	• Effective usage of MoC processes across the organization. • Increased conformance to MoC standards across the organization. • Maintain and continuously improve standards by regularly engaging the network. • Timely response to questions/requests from the business.	• Monitor the current state of MoC across the organization. • Define the MoC metrics for improving MoC performance. • Drive the improvement of the standard, definitions, and audit protocol documents. • Seek out best practices from across the organization and industry for organizational sharing. • Drive the type and quality of MoC training required. • Participate in PSM audits relating to MoC as needed. • Interface and collaborate with other networks.	• Audit and self-assessment of performance. • Improvements in performance on corporate and local MoC KPIs and metrics. • Reduction in turnaround time for support/responses to queries from business and CoPs. • Demonstrated behaviors endorsing the use and application of the MoC standard and its intent.	• Results oriented. • Team composition. • BU approval and support of members' commitment. • Effective steering/facilitation. • Excellent communication and visibility. • Clear performance measures. • Membership and performance of the members in their personal goals and scorecards. • Best-in-class infrastructure to support efforts. • Collaborating with other networks.

- Build competencies required to perform MoC in the communities of practice (CoPs).
- Reduce number or severity of process safety incidents related to MoC.

In Scope	Out of Scope	Deliverables	Members	Meetings
• Assist in revision and continuous improvement of MoC standards. • Development and enhancement of training material (generic—not site specific). • Engagement with the network covering all business units/areas. • Engagement with other networks as appropriate. • Key stakeholders for MoC (IT) tool development/sustainment. • Inclusion of management system requirements for management of change.	• MoC-people. • Site implementation of MoC or other site specific initiatives. • Site selection of MoC personnel. • MoC IT tool implementation and sustainment.	• An up-to-date MoC standard, audit protocols, PSM definitions. • Up-to-date training materials. • Monthly stewardship summaries of activity and results.	• Sponsor. • Team lead. • Team members.	• Notionally 12 meetings per year. • One face-to-face annually. • One virtual meeting per year (1–3 hours). • Initial face-to-face meeting of all members (2 days). • Monthly update to steering team.

Source: Lutchman, C. et al., *Process Safety Management: Leveraging Networks and Communities of Practice for Continuous Improvement*, Taylor & Francis, CRC Press, Boca Raton, FL, 2013.

network forum on a minimum quarterly basis. This enables collaboration and sharing among the network leaders to ensure consistent practices, processes, tools. and templates are being utilized across the networks.

This quarterly network leaders forum provides opportunities for the following:

- Sharing updates
- Sharing successes that may be applicable to other networks
- Identifying obstacles and roadblocks that are common to several networks
- Finding solutions to common obstacles and roadblocks
- Network motivation
- Sharing processes and tools to enable commonality and standardization

These network integration and forum opportunities are great for celebrating wins while at the same time motivating networks to higher levels of performance.

12.6 Networks Governance and Leadership

Network Governance

The principal role of network governance is to enable the work and efforts of each network. Network governance is performed by a network steering team that is comprised of the senior leaders of the organization. This team seeks to enable all networks by providing support in the following ways:

- Removing obstacles
- Promoting visibility and recognition
- Ensuring alignment across multiple networks
- Screening, selecting, approving, and activating—function as a filter mechanism in determining the appropriateness of new networks
- Recommending corporate resources
- Stewarding performance and serving as a conduit to senior corporate leaders
- Facilitating the celebration of wins
- Mentoring/coaching the network leaders

Staffed with four to six senior organizational leaders charged with decision-making authority, the network steering team forms the interface between corporate leadership, business unit leaders (who are impacted by the outputs of networks), and the network leader and core team. This group

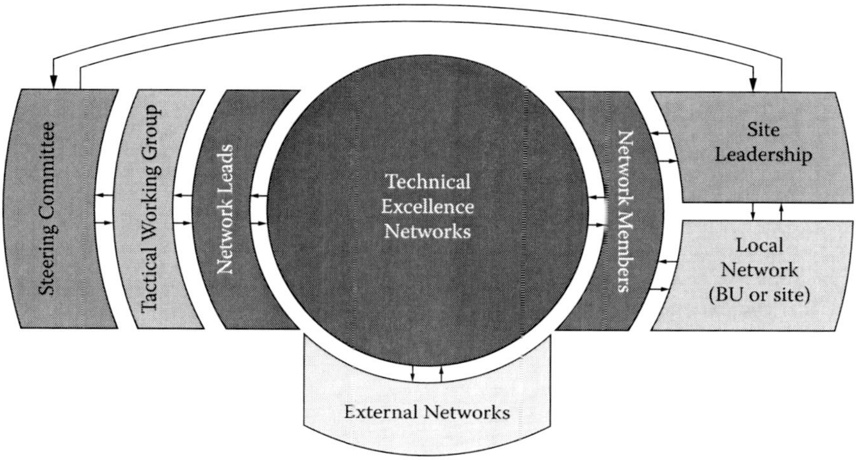

FIGURE 12.2
Sample network governance model. (From Suncor Energy, Suncor Operational Excellence, Internal Reference Document, 2014, retrieved February 4, 2014, from Suncor Energy.)

of highly experienced and competent leadership professionals is knowledgeable in technical disciplines as well as leadership. They are champions of organizational change management. Figure 12.2 provides a simplified version of a network governance model that highlights the interface among key stakeholder groups.

Stewardship of networks is best achieved through a simplified process that provides the steering team a quick snapshot of the network support requirements, achievements, outlook, and status updates on an acceptable frequency. Table 12.4 provides a sample MoC stewardship and network reporting template.

12.6.1 Network Leadership

Leading a network is critical to the success of the network. Since the core team is comprised of experienced, mature workers, failing to apply the right leadership skills and behaviors may result in a dysfunctional network. Nonauthoritative, transformational leadership skills and behaviors are most appropriate for leading networks.

Lutchman et al. (2012, p. 70) suggested the key leadership behaviors and attributes of the transformational leader are as follows:

- Will create an organizational environment that encourages creativity, innovation, proactivity, responsibility, and excellence
- Has moral authority derived from trustworthiness, competence, sense of fairness, sincerity of purpose, and personality

- Will create a shared vision; promote involvement, consultation, and participation
- Leads through periods of challenges, ambiguity, and intense competition or high growth periods
- Promotes intellectual stimulation
- Usually considers individual capabilities of employees
- Is willing to take risks and generate and manage change
- Leads across cultures and international borders
- Builds strong teams while focusing on macro-management
- Is charismatic and motivates workers to strong performance

Being able to apply many of these skills while leading a network is a key success factor for delivering a strong network performance.

TABLE 12.4

Sample MoC Network Stewardship Report

Network – MoC	**Date: September 14, 2012**
Lead: (Insert Network Lead's Name Here)	
Overall Status	On target
Sub Initiative Updates(s)	o Collaboration site completed, tested, and activated across all business units. o Community of practice members identified and initial communication sent to members. o No clear agreement on definitions of MoC-E/MoC-NE. Continue to work with SMEs for agreement. o MoC practices and procedures from three of five BUs received and under review by core team for organization wide application.
Next Period Outlook	o Monitor use of collaboration site and work with community of practice to maximize value of collaboration. o Obtain practices and procedures form remaining business units for evaluation. o Obtain industry MoC practice and procedures for best practice comparison. o Conduct face-to-face core team review of practices and procedures to identify MoC best practices. o Determine 3–5 high value KPIs for MoC stewardship and performance management. o Select and roll-out MoC best practices across all BUs.

Support Requirements	
Steering Team	o Need help in securing SME from BU 1. o Support required in getting BU 2 leadership onboard in use of collaboration site within BU.
Issues and/or Challenges	o Difficulties in accessing SMEs given current work loads. o Leveraging technology for communication and meetings is working to an extent. Require a face-to-face meeting to resolve critical issues and decisions.

Network Successes
o Rollout of collaboration site with good acceptance and use across BUs for sharing MoC information. o Community of practice team members identified and approved. o SMEs identified and responding to the cause. Motivated team of personnel working together. o All members of the core team in place. o Charter finalized and approved by steering team.

Network Participation	
Meeting attendance	
% of network absenteeism	70%
# of meetings cancelled	2
# of network vacancies	Zero

Source: Lutchman, C. et al., *Process Safety Management: Leveraging Networks and Communities of Practice for Continuous Improvement*, Taylor & Francis, CRC Press, Boca Raton, FL, 2013.

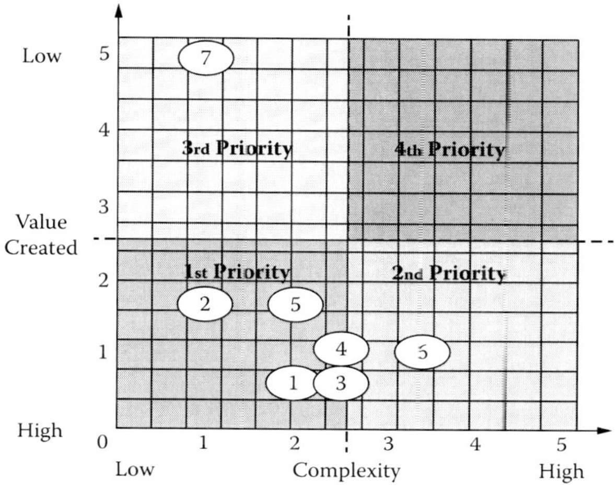

Priority	Opportunity	Complexity 1=Low 5=High	Value Created 1=High 5=Low
1	Get prequalifiaction rolled out right	2.0	0.5
2	Align contract language with PSM/OMS requirements	1.0	1.5
3	Standardized Computer Based Contractor Orientation (3 levels)	2.5	0.5
4	Simplify CM101 and make BU/BA specific (workshop with tools)	2.5	1.0
5	CM 101 support to BU in rollout of Contractor Safety	2.0	2.0
6	Provide access to organization's work related information (procedures, policies, processes) to all contractors	3.5	1.0
7	Update the Contractor Safety Standard based on improvement opportunities identified	1.0	5.0

FIGURE 12.3
Application of the opportunity matrix in prioritizing work of a contractor safety management network. (From Lutchman, C. et al., *Process Safety Management: Leveraging Networks and Communities of Practice for Continuous Improvement*, Taylor & Francis, CRC Press, Boca Raton, FL, 2013. Copyright © 2012 Safety Erudite, Inc.)

Network leaders should also have the ability to prioritize work effectively for value maximization. Figure 12.3 provides a sample opportunity matrix that has been applied in prioritizing the work of a contractor safety management network. A similar process for prioritizing the rollout of initiatives and knowledge generated by networks is provided in the value–effort relationship matrix demonstrated in Figure 12.4.

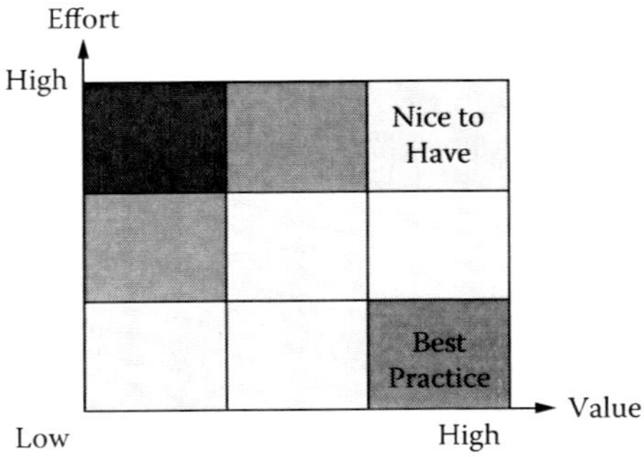

FIGURE 12.4
Value–effort matrix for acting on knowledge and learnings generated. (From Lutchman, C. et al., *Process Safety Management: Leveraging Networks and Communities of Practice for Continuous Improvement*, Taylor & Francis, CRC Press, Boca Raton, FL, 2013. Copyright © 2012 Safety Erudite, Inc.)

12.7 Performance Results

As with any business management project or initiative, performance results are an integral component of the degree of success of the project. Performance results of networks are also important in assessing the sustainability of networks. Perhaps the best measure for evaluating the performance of a network is to reduce its value contributions to its simplest form: dollars and cents.

In many instances, however, it may not be easy to reduce network performance results and value contributions to a simple cost-benefit ratio, or even to financial terms, such as net present value (NPV) or internal rates of return (IRRs), because of difficulties in allocating a dollar value to intangible benefits. Among the more common measures for evaluating the performance of networks are the following:

- Value measure—dollars cost–benefit ratio
- Operating and financial risk reduction
- Declining trends in the number of tier 1 and tier 2 incidents or near misses
- Improvements in operating reliability and availability
- Building organizational and technical discipline competency
- Defining and adopting new best practices
- Capturing and sharing knowledge

12.8 Conclusion

As discussed earlier in this chapter, studies have shown that 90 to 95% of workplace incidents are avoidable and 80 to 85% are repeated. In view of this, questions must be asked regarding our ability to learn internally and from external sources. Is it that organizations are failing to learn proactively from internal and external sources because they are unable to do so? Do they not understand how to do so? Do they not see the need to learn?

The authors are convinced that most organizations are very much in search of simple and effective methods for generating and capturing knowledge. These organizations are also seeking methods for transferring knowledge generated into applied knowledge in the field to enhance business performance. Organizations don't have memories, people do. This universal truth should drive organizations to actively find effective methods for capturing, retaining, and sustaining knowledge. Employees will retire, change jobs, move, and forget!

Formal networks have the potential to provide real solutions to the learning and sustainment challenge faced by organizations. This opportunity is severely underutilized by organizations in the quest for improvements in reliability, reduction in the number of incidents, improvements in operating efficiency and performance, and value maximization. The chapter explored networks as a viable model for generating PSM knowledge from both internal and external sources, and transforming this knowledge into applied practices at the front line.

This proven practice has been in effect in many leading oil and gas organizations, producing strong PSM performance in many of these organizations. While the concept of formal networks is relatively new, informal networks have been in existence for decades, producing great organizational performance. Formal networks provide organizations opportunities to move from great to a *sustained excellence* business performance. The knowledge provided in this section builds on already published work by Lutchman et al. (2013), which addresses continuous improvements in PSM by leveraging networks and communities of practices.

References

Hopkins, A. (2012). Disastrous decisions: The human and organizational causes of the Gulf of Mexico blowout. *Journal of World Energy Law and Business*, 5(4). Retrieved January 11, 2014, from EBSCOHost database.

Kletz, T. (2001). *Learning from accidents.* Gulf Professional Publishing, an imprint of Butterworth Heinemann. Retrieved January 11, 2014, from http://books.google.ca/books?hl=en&lr=&id=2fUgzeXWDcgC&oi=fnd&pg=PR3&d

q=learning+from+accidents+kletz&ots=bFffw4NCjZ&sig=Hh0krTLPltg TU2FMjf-G6L_Lu3o#v=onepage&q=learning%20from%20accidents%20 kletz&f=false.

Lutchman, C. (2012). How to go from lesson to learned: PSM from engineering to operations. Presented at the 8th Global Congress on Process Safety, Houston, TX, April 1–4.

Lutchman, C., Maharaj, R., and Ghanem, W. (2012). *Safety management: A comprehensive approach to developing a sustainable system*. Taylor & Francis, CRC Press, Boca Raton, FL.

Lutchman, C., Evans, D., Maharaj, R., and Sharma, R. (2013). *Process safety management: Leveraging networks and communities of practice for continuous improvement*. Taylor & Francis, CRC Press, Boca Raton, FL.

Raleigh, P. (2013). Learning the lessons. *Process Engineering*, 94(2), 03701859. Retrieved January 11, 2014, from EBSCOHost database.

Safety Erudite. (2014). CCS: Collect, standardize and share. Our high value low effort transformational approach to improving safety in your workplace through Shared knowledge. Retrieved December 27, 2014, from https://www.safetyerudite.com/.

Suncor Energy. (2014). Suncor operational excellence. Internal reference document. Retrieved February 4, 2014, from Suncor Energy.

Section III

Workshop Fundamentals of an Operational Excellence Management System

Note: Readers can download an electronic copy of the workshop slides at a nominal fee from www.safetyerudite.com.

7 Fundamentals of an Operationally Excellent Management System®

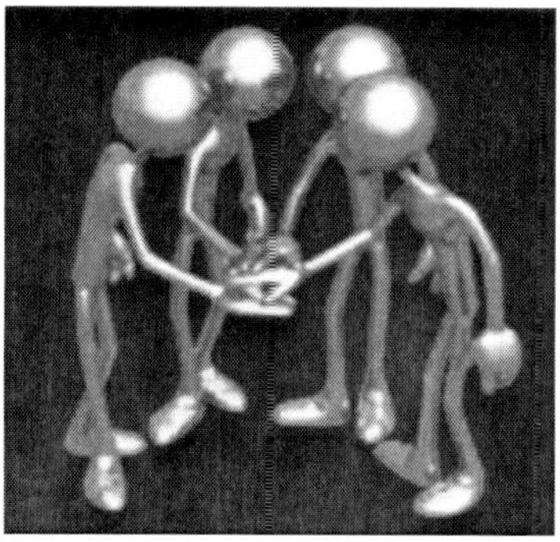

A three day workshop designed to provide all leaders and personnel an overview of the *People, Processes,* and *Facilities* requirements for Operational Discipline and Excellence.

A Product of Safety Erudite Inc.

By Dr. Chit Lutchman
CSP, CRSP, 1st Class Power Engineer

Fundamentals of an Operationally Excellent Management System®

Operations Discipline and Operations Excellence Lead to World Class Performance – and It Begins with You ...

_______________________________ _______________

Your Name *Date*

Sometimes we become a big company and we are required to learn to behave like a big company. – Chris Aman (2011)

Important Notice

Fundamentals of an Operationally Excellent Management System is a facilitator-led workshop. Facilitators must be Safety Erudite Inc. certified. Recipients of this workshop are bound by the terms and conditions of the governing license between Safety Erudite Inc. and the company within which this workshop is executed.

This workshop was created to provide future leaders of organizations an overview of the requirements of an Operationally Excellent Management System.

Contents

Let's light the fire in the hearts of our people as opposed to under their feet. – Anonymous (2012)

Discussion Topics

- OEMS – A definition
- Benefits of an OEMS
  - Employer of choice – high employee morale
  - High productivity and profitability
  - Goodwill for environment and socially responsible behaviors
  - More effective knowledge management/transfer and consistency in decision-making
  - Improved customer, contractor, and supplier relationship
  - Peer and industry recognition
- People, Processes, and Systems, and Facilities and Technology
- Who needs to know about it?
  - All workers, starting first with leaders
  - Contractors, service providers, and materials suppliers
  - All stakeholders

Give a man a fish and feed him for a day. Teach a man to fish and feed him for a lifetime.

– Lao Tzu

Introduction

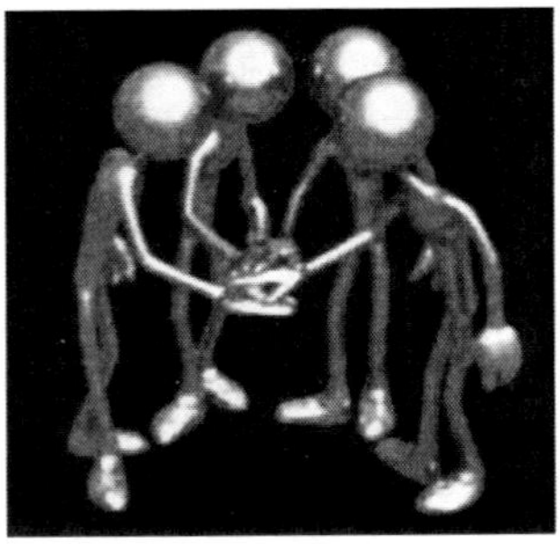

In this section you will gain insights into

1. What constitutes an Operationally Excellent Management System
2. What are the benefits of an Operationally Excellent Management System
3. Why it is important for you to know about Operationally Excellent Management Systems

Discussion Topics

- OEMS – A definition
- Benefits of an OEMS
    - Employer of choice – high employee morale
    - High productivity and profitability
    - Goodwill for environment and socially responsible behaviors
    - More effective knowledge management/transfer and consistency in decision-making
    - Improved customer, contractor, and supplier relationship
    - Peer and industry recognition
- People, Processes, and Systems, and Facilities and Technology
- Who needs to know about it?
    - All workers, starting first with leaders
    - Contractors, service providers, and materials suppliers
    - All stakeholders

Give a man a fish and feed him for a day. Teach a man to fish and feed him for a lifetime.

– Lao Tzu

OEMS – A Definition

An Operationally Excellent Management System is one that

- Develops its people to high levels of discipline and competence
- Is supported by adequate tools and processes, adjusts to the continually changing business environment, and is guided by policies, standards, and procedures for ensuring the integrity of its assets
- Provides standardization for global operations and business practices
- Prioritizes the management of environment, health and safety (EH&S), and continually improves the reliability and efficiency of business performance
- In the presence of strong leadership commitment, caters for adopting best practices and standards that help in the delivery of world-class performance and sustained value maximization
- Demonstrates sustained exceptional performance and is noted for few incidents and preservation of the environment

Operational excellence is not something separate from our business; it is how we must run our business to achieve our vision of success.

– Chevron

Benefits of an OEMS

- Employer of choice – high employee morale
- High productivity and profitability – standardization and elimination of duplicated efforts
- Goodwill for environment and socially responsible behaviors
- More effective knowledge management/transfer and consistency in decision-making
- Improved customer, contractor, and supplier relationship
- Peer and industry recognition

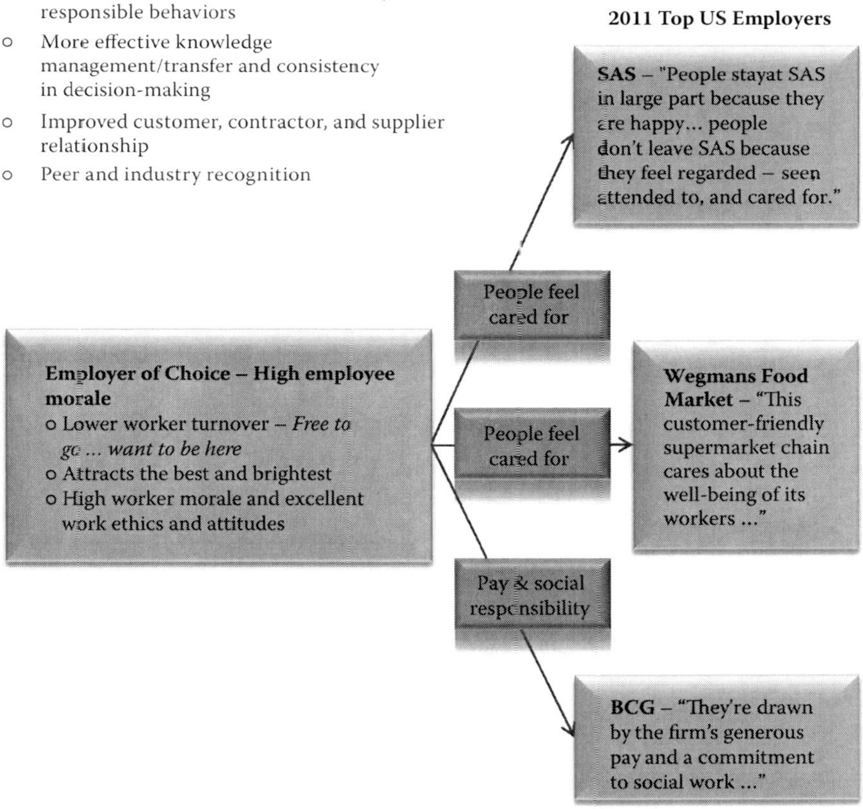

Benefits of an OEMS

High productivity and profitability – standardization and elimination of duplicated efforts. Doing it right once only.

- Fewer incidents
- High operating reliability
- Greater operating efficiency
- More efficient use of scarce resources
- Maximized value creation

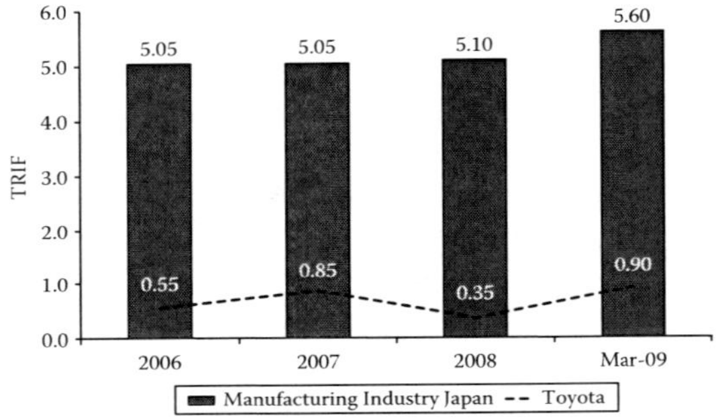

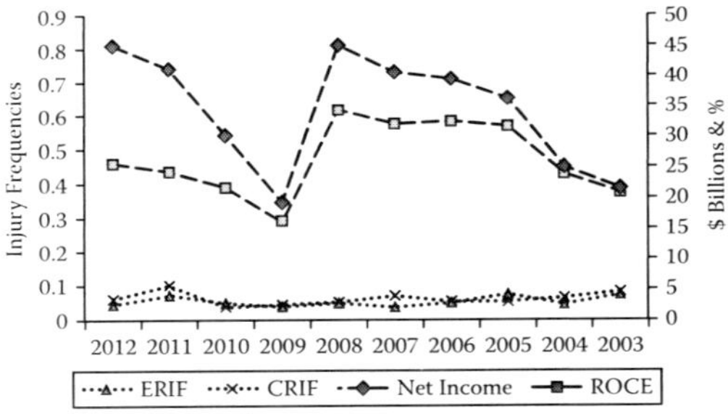

Exxon Mobil – 2013

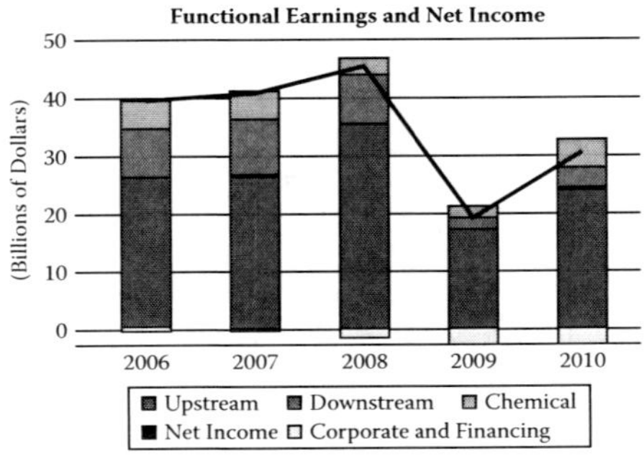

Benefits of an OEMS

Environment and Social Responsibility

- Minimizes impact to the environment
- Returns to society in the forms of employment, community development, and disaster support
- Excellent corporate goodwill and image

Consequences	Mitigation
○ Greenhouse gas emissions – global warming ○ Deforestation ○ Stream and river contamination ○ Floods and erosion ○ Habitat destruction	○ Flaring and burner management control ○ Land use optimization ○ Water recycle programs ○ Reforestation ○ Managed land use

There is a sufficiency in the world for man's need but not for man's greed.
— Mohandas K. Gandhi

Benefits of an OEMS

More effective knowledge management/transfer and consistency in decision making

- Creates a learning culture
- Fewer repeat incidents
- Better and more timely value-added decisions
- Quality knowledge shared
- Applied knowledge

Model for Shared Learning

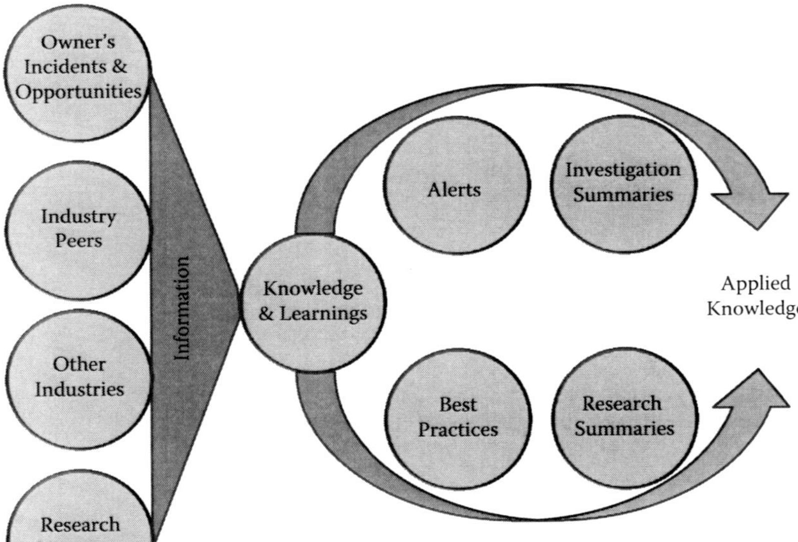

Learning is an active process. We learn by doing. Only knowledge that is used sticks in your mind.
– Dale Carnegie

Benefits of an OEMS

Improved Customer, Contractor, and Supplier Relationship

- Greater efficiency and performance from shared values
- Contractors and suppliers know what is expected of them when working at your company
- Less rework required
- On budget and on schedule delivery of projects

What Contractors and Suppliers Want	What Owners Need to Do
○ Engagement and involvement ○ To do a great job ○ To hear that they did a great job ○ To know where and how they can improve ○ To be paid on time ○ More work	○ Provide full access to relevant information for assigned work ○ Provide clear scopes of work ○ Provide timely and accurate feedback ○ Manage payment ○ Conduct information/feedback forums

The purpose of business is to create and keep a customer.
– Peter Drucker

Benefits of an OEMS

Peer and Industry Recognition

- Know-how and competence
- Best practices
- Goodwill

Organizational Performance

- Standardized processes
- Operational discipline
- Strong reliability
- Organized and planned performance
- Output and financial performance
- Excellent environmental performance
- Performance development
- Motivated workforce
- Little turnover

The purpose of business is to create and keep a customer.
– Peter Drucker

Elements of an OEMS: People, Processes, & Systems, and Facilities & Technology

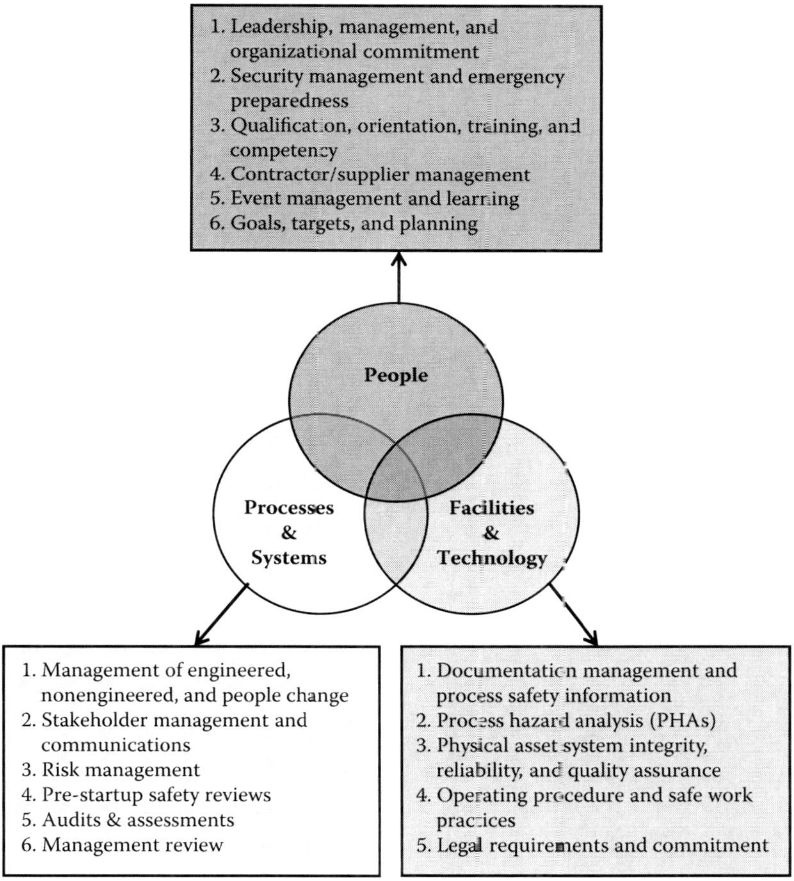

Who needs to know about it?

- All workers, starting firstly with leaders
- Contractors, service providers, and materials suppliers
- All stakeholders

All Workers Starting Firstly with Leaders

- Vision
- Values
- Culture

Contractors, Service Providers, and Materials Suppliers

- Owner's vision
- Owner's values
- Owner's culture

Fundamental 1

Leadership Commitment and Motivation

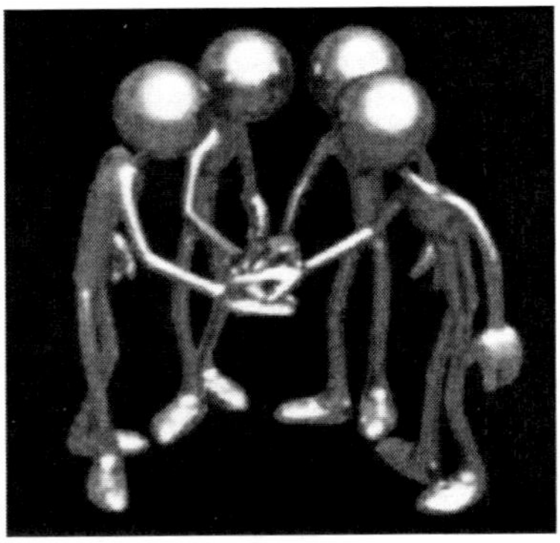

In this section you will gain insights into

1. The Evolving Leadership Environment
2. The Leadership Conundrum – Gen X & Gen Y
3. Leadership Behaviors for Establishing and Leading an Operationally Excellent Management System

Discussion Topics

- What is leadership? A definition
- Evolving leadership environment
- The workplace conundrum – Gen X & Gen Y
- Leadership styles and behaviors
    - Autocratic
    - Democratic
    - Servant
    - Situational
    - Transformational
- Leadership at the frontline
- Situ-transformational leadership for establishing and leading an Operationally Excellent Management System

Leadership: the art of getting someone else to do something you want done because he wants to do it.

– Dwight D. Eisenhower

What Is Leadership? A Definition

Some Suggestions

- Simplest form: *leadership is one's ability to influence the behaviors of another.*
- Leaders create a compelling sense of direction for followers and motivate them to performance levels that will not generally occur in the absence of the leader's influence.
- Leadership is about creating an organizational environment that encourages worker creativity, innovation, proactivity, responsibility, and excellence.
- True leadership begins with understanding your own leadership style and its impact on your followers.

Management is doing things right; leadership is doing the right things.

– Peter Drucker

The Role of Leadership Is to

- Provide leadership and direction
- Provide prioritization and resources
- Establish written standards and supporting procedures and defines roles and responsibilities
- Set goals, objectives, and expectations for worker performance
- Establish accountability for performance against goals and objectives
- Establish and steward performance KPIs
- Audit the EH&S management process
- Provide oversight of work and performance

Management is doing things right;
leadership is doing the right things.

– Peter Drucker

Industry Trends Impacting Leadership

- A more educated and knowledgeable workforce
- More women in the workforce
- Shorter tenure in the workforce
- Declining union membership and influence of unions
- Increasing numbers of Gen Y in the workforce
- Information revolution
- Thinner margins
- Scarce and declining resources
- Political instability
- A global workspace
- Continuously evolving technology
- Greater stakeholders' expectations

Source: US Department of Labor, Bureau of Labor Statistics, 2006

Industry Trends Impacting Leadership

A More Educated and Knowledgeable Workforce

- Graduates with degrees almost doubled since the early 1970s
- Fewer unskilled workers with less than a high school diploma
- More difficult to attract and retain given competition for skilled and highly trained workers

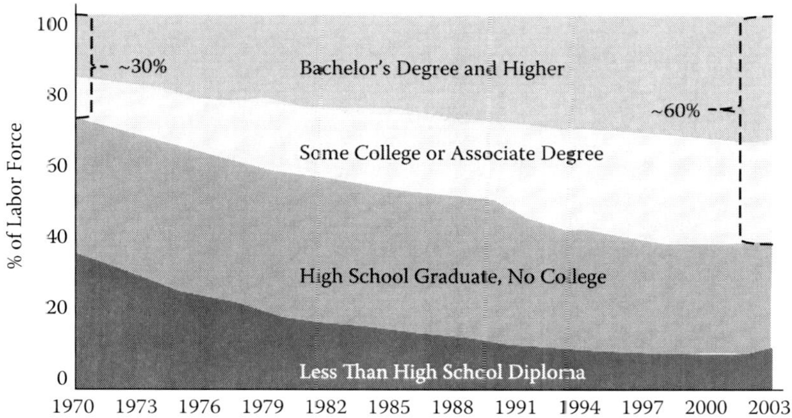

Source: US Department of Labor, Bureau of Labor Statistics, 2006

Industry Trends Impacting Leadership

More Women in the Workforce – Workplace and Hygiene Factors Considerations

- o Growing workforce needs
- o Well-educated and trained women
- o Nontraditional jobs and work environments
- o Ergonomics considerations
- o People skills

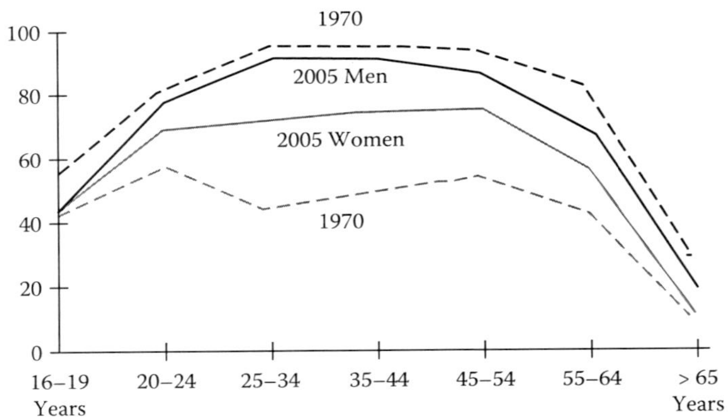

Source: United States Bureau of Labor Statistics, 2009.

Industry Trends Impacting Leadership

Shorter Tenure in the Workforce – Retention Concerns

- Greater choices
- Two-income families… retiring faster
- More educated workforce with ability to command higher wages

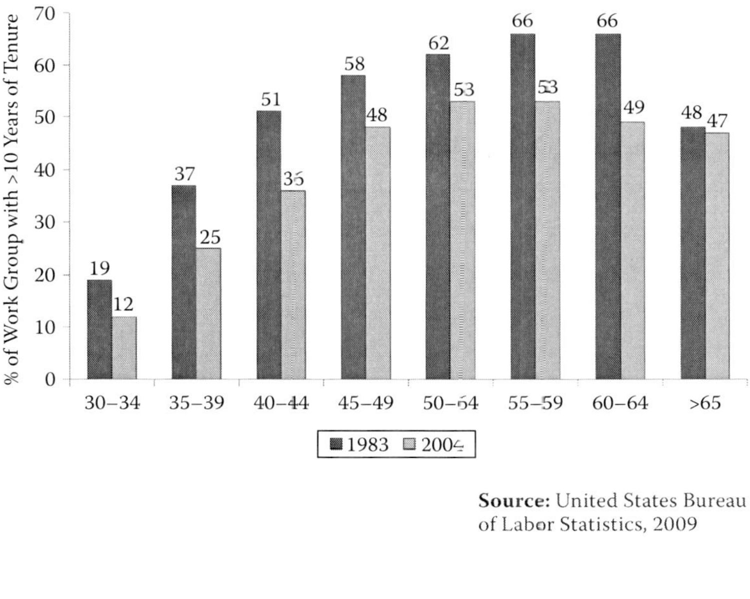

Source: United States Bureau of Labor Statistics, 2009

Industry Trends Impacting Leadership

Declining Union Membership and Influence of Unions

- Primarily focused on frontline skilled trades people
- Automation
- Businesses have improved working conditions – Hygiene and Health and Safety factors considerably (early focus of unions)

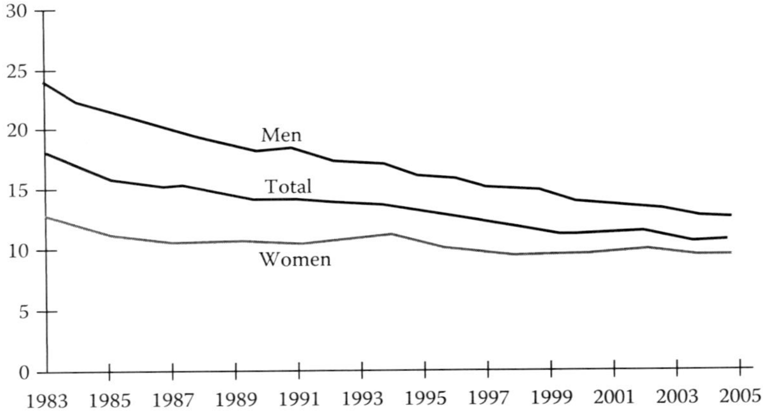

Source: United States Bureau of Labor Statistics, 2009.

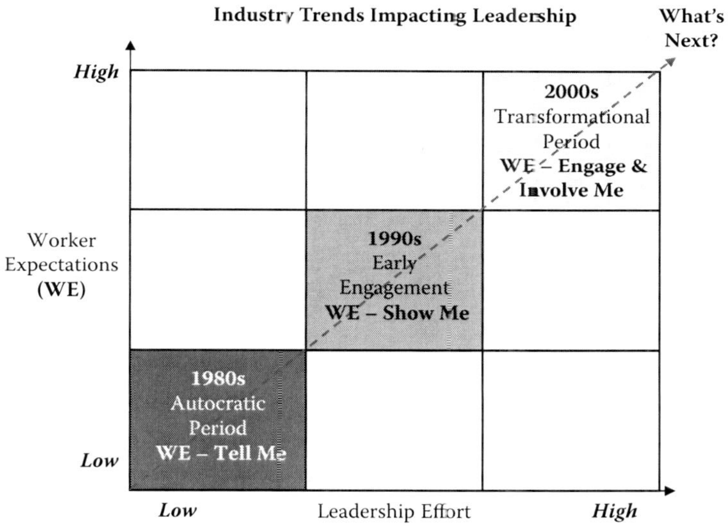

Source: Lutchman, 2010

Leadership Expectations of the Future

- What does Gen Y expect from the workplace ...

  - Master of my destiny: _________________
  - Little field experience – Train me: _________________
  - Flexible hours and locations: _________________
  - I want to grow quickly and take charge: _________________
  - I want to take charge: _________________

____

____

____

____

____

____

____

____

____

Leadership Expectations of the Future

- Increasing numbers of Gen Y in the workplace ...
    - ➢ Master of my destiny – INVOLVE ME
    - ➢ Little field experience – PROTECT ME
    - ➢ Flexible hours and locations – REMOTE ME
    - ➢ I want to grow quickly and take charge – PROMOTE ME
    - ➢ I want to take charge – EMPOWER ME

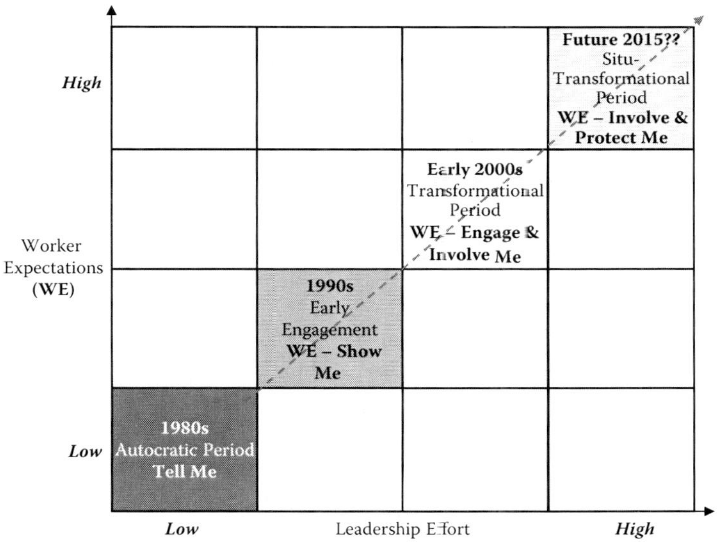

Generation Y: Who Are They?

- Workers entering the workforce generally <35 years old
- Computer literate and eager to please
- Want a safe workplace and eager for safety training in the workplace
- Strong ability to multitask
- Willing to embrace change
- Strong ability to absorb information – and willing to ask questions when not sure
- Demand more from trainers
- Demand training from multimedia, one-on-one, and print materials
- Will change jobs 4–5 times in first 10 years
- Have high expectations of employers, and they must be known and catered for
- Employers viewed upon as a resource pool for relationships, learning, and opportunities to excel and to be rewarded
- Utilize social media such as Twitter and Facebook
- Dislike authoritarian leadership and management styles

Source: Lutchman, 2010

Leadership Expectations of the Future

Retaining Generation Y Workers

- Keep them safe and motivated
- Involve and engage them in work decisions
- Leverage information technology to cater to their learning needs (computers/videos)
- Maintain dynamic and knowledgeable training providers
- Be flexible in working hours and training methods
- Demonstrate and replicate ... use role play to simulate situations

Source: Lutchman, 2010

Leadership Challenges in the Work Environment

- Competition for scarce resources
- Thin margins
- Highly regulated environment
- Global competition on an uneven field
- Mixed global workforce
  - Migrant labor
  - Temporary foreign workers (TFWs)
  - Retention
  - Language differences
  - Cultural differences
- Free information flow at lightning speed
  - Gulf of Mexico spill – 2010
  - Fukushima Daiichi nuclear disaster – Japan 2011

Source: Lutchman, 2010

Gen X … Gen Y Challenge

Gen X

- Prefer print materials in learning
- Tremendous field experience and may therefore be more set in ways
- Shy of information and computer technology
- Change only if absolutely necessary
- Can be delegated to

Gen Y

- More effective with computers and information technology learning
- Lack field experience which encourages them to trust leaders
- Shorter attention spans and require clear and precise messaging
- Need trendy PPE
- Need change and thrive in changing environments
- Need to be shown how, what to do
- Risk takers

The best way to find out if you can trust somebody is to trust them.

– Ernest Hemingway

Improved Customer, Contractor, and Supplier Relationship

- Greater efficiency and performance from shared values
- Contractors and suppliers know what is expected of them when working at your company
- Less rework required
- On-budget and on-schedule delivery of projects

What Contractors and Suppliers Want	What Owners Need to Do
o Engagement and involvement o To do a great job o To hear that they did a great job o To know where and how they can improve o To be paid on time o More work	o Provide full access to relevant information for assigned work o Provide clear scopes of work o Provide timely and accurate feedback o Manage payment o Conduct information / feedback forums o **Have an end-to-end contractor management process**

The purpose of business is to create and keep a customer.
– Peter Drucker

What We Say and What Others Hear

True Message	Organizational Level	Conflicting Messages
Safety is Our Top Priority	**CEO & Senior Leadership**	o Market and shareholders reward profits o CEO loses job when profits are lower than peers but safety record is better o Not willing to recognize a lag in profits between investments in safety and long-term sustainable profitability
Safety before Production	**Middle Managers**	o Production rewarded over safety o Downtime and outage frowned upon – correcting safety deficiencies requires downtime and outage o Preventive maintenance often deferred based on pricing and profit drivers o Demotions and job loss as an outcome of lower than peer production performance o Weak recognition of a lag between sustained productivity and investment in safety
Safety First	**Frontline Supervisors and Workers**	o Frontline supervisors poorly trained o Senior and frontline leaders talk safety but do not demonstrate the behavior … why should I? o Getting the job done quickly o Nondetected shortcuts rewarded
Safety First	**Contractors**	o Bid process rewards lowest prices – safety budgets generally the first to be chopped to be competitive o Getting the job done quickly rewarded o Contractors often provided the dirtiest, most difficult, and most dangerous jobs o Punitive consequences for reporting incidents and near misses drives reporting underground leading to lost learning opportunities

Source: Lutchman, Maharaj, & Ghanem (2012)

The ABCD of Trust

ABLE – Demonstrate competence:

- Produce results
- Make things happen
- Know the organization/set people up for success

BELIEVABLE – Act with integrity ... be credible:

- Be honest in dealing with people/be fair/equitable/ consistent/respectful
- Values-driven behaviors "reassures employees that they can rely on their leaders" (p. 2)

CONNECTED – Demonstrate genuine care and empathy for people:

- Understand and act on worker needs/listen/share information/be a real person
- When leaders share a little bit about themselves, it makes them approachable

DEPENDABLE – Follow through on commitments:

- Say what you will do and do what you say you will
- Be responsive to the needs of others
- Being organized reassures followers

Source: The Ken Blanchard Group of Companies (2010)

Trust – You Are in Charge ...

Erodes Trust

- Lack of communication
- Being dishonest
- Breaking confidentiality
- Taking credit for others' work
- Unethical behaviors
- Nepotism/favoritism

Builds Trust

- Giving credit
- Listening
- Setting clear goals
- Being honest
- Following through on commitments
- Caring for your people

Building Organizational Trust

Builds Trust

- Demonstrate trust in your people
- Share information
- Tell it straight/never lie
- Create a win-win environment
- Provide feedback
- Resolve concerns head-on
- Admit mistakes when you make them ... we are all humans
- Recognize positive behaviors in a timely manner

The best way to find out if you can trust somebody is to trust them.
– Ernest Hemingway

Alignment of Annual Goals & Objectives to Achieve Strategic Objectives

Develop and Communicate Strategic Goals and Objectives	**Corporate Leadership**
Develop and Communicate Annual or Short-term Goals and Objectives	**Site Leadership** • Ensures annual goals are aligned with corporate goals and objectives • Develop tactical strategies for achieving corporate goals • Provide oversight and stewardship (e.g., trend analysis) to identify improvement opportunities and resource prioritization
Participate in developing and communicating short-term Goals and Objectives	**Middle Managers** • Develop and steward tools for collecting data and measuring progress relative to short-term goals and objectives • Identify and close gaps that may prevent achieving short-term goals
Supervise and manage work safely to ensure short-term Goals and Objectives are met	**Frontline Supervisors and Workers** • Ensure work is conducted in a way so as to ensure short-term goals and objectives are met • Ensure work-plan contain tactical measures and strategies to achieve short-term goals • Promote continuous improvements in safety and worker performance
Supervise and manage work safely to ensure short-term Goals and Objectives are met	**Contractors** • Ensure work is conducted in a way so as to ensure short-term goals and objectives are met • Ensure work-plan contains tactical measures and strategies to achieve short-term goals • Promote continuous improvements in safety and worker performance

Source: Lutchman, Maharaj, & Ghanem (2012)

Alignment of Annual Goals & Objectives to Achieve Strategic Objectives

Develop and maintain corporate standard	**Corporate Leadership**
Develop and maintain supporting procedures	**Site Leadership** • Define and approve template for procedure • Ensures procedure alignment with standard • Identify site roles and responsibilities in applying the procedure • Drives compliance in use of procedures • Recommends improvements to the standard as required
Enforcement in use of procedures	**Middle Managers** • Ensures all personnel are trained and competent in the use of the procedure • Ensures procedures are accessible to all users • Verifies procedures are correct and upgrades when deficient in meeting safe working conditions
Consistent use of procedures	**Frontline Supervisors and Workers** • Provide oversight in the use of established procedures • Ensure all workers are trained and competent in the use of the procedures • Promote continuous improvements in the application of the procedure
Consistent use of procedures	**Contractors** • Provide oversight in the use of established procedures • Ensure all workers are trained and competent in the use of the procedures • Promote continuous improvements in the application of the procedure

Source: Lutchman, Maharaj, & Ghanem (2012)

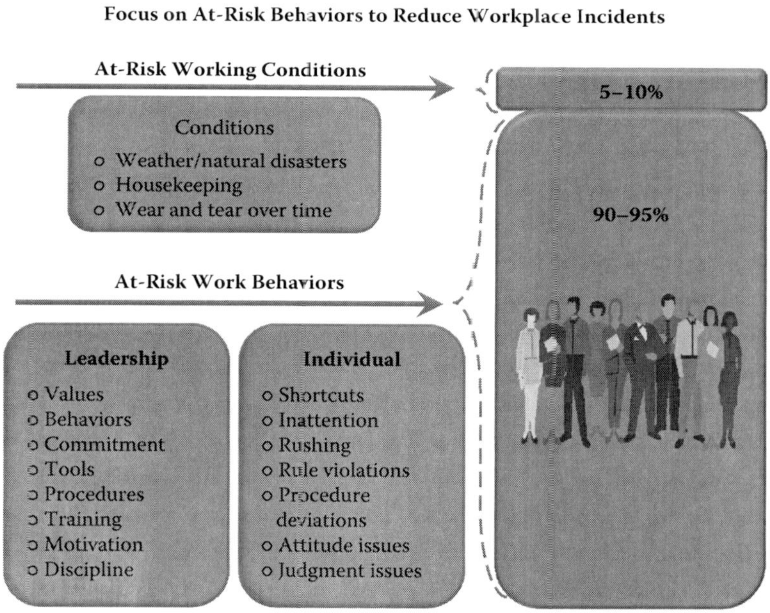
Focus on At-Risk Behaviors to Reduce Workplace Incidents
At-Risk Working Conditions
Conditions
o Weather/natural disasters
o Housekeeping
o Wear and tear over time
At-Risk Work Behaviors
Leadership
o Values
o Behaviors
o Commitment
o Tools
o Procedures
o Training
o Motivation
o Discipline
Individual
o Shortcuts
o Inattention
o Rushing
o Rule violations
o Procedure deviations
o Attitude issues
o Judgment issues
5–10%
90–95%

Perceptions and Stereotyping

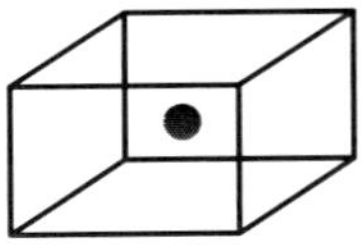

On which face of the cube is the dot?

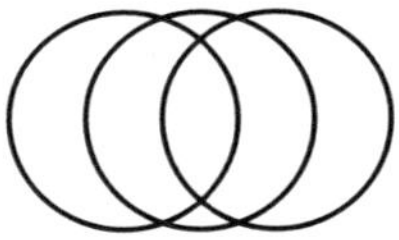

Which end of the circular tube are you looking through?

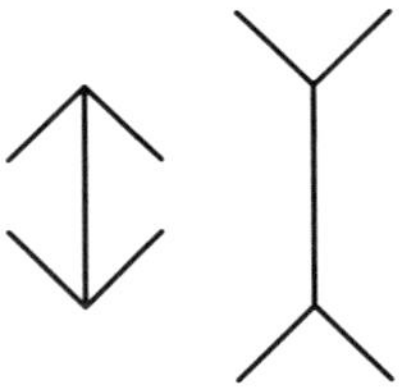

Which vertical line is longer?

Are the lines parallel?

Autocratic Leaders

Strengths	**Weaknesses**
<ul><li>Looks after the safety of equipment and people</li><li>Returns structure to a poorly managed area</li><li>Makes decisions quickly</li><li>Credited for running a tight ship</li></ul>	<ul><li>Relies on threats and punishment to influence employees</li><li>Will not trust employees</li><li>Disallows employee input</li><li>Limits creativity of workers</li><li>Creates the Theory X employee: the lazy, unmotivated, here-for-the-paycheck employee</li><li>An experienced and competent leader is required</li></ul>

Source: Lutchman, Maharaj, & Ghanem (2012)

Democratic Leaders

Strengths	**Weaknesses**
<ul><li>Can maintain high quality and productivity over long periods of time</li><li>Employees like the trust they receive and respond with cooperation, team spirit, and high morale</li><li>Often recognizes and encourages achievements</li><li>Generally, encourages employees to grow on the job and be promoted</li></ul>	<ul><li>Mistakes, when they occur, can be quite costly</li><li>The decision-making process can be very lengthy as leaders try to get everyone on board</li><li>The lengthy decision-making process adds to the overall cost of doing business</li><li>Opportunities may be lost while leaders engage other stakeholders to gain support for decisions</li></ul>

Source: Lutchman, Maharaj, & Ghanem (2012)

Servant Leaders

Strengths	**Weaknesses**
o Promote empowerment and mutual trust o Possess skills of listening, empathy, healing, awareness, persuasion, conceptualization, foresight, stewardship, and commitment to the growth of people o Skilled in consensus-making, ethical decision-making, and conflict resolution	o Cannot perform in volatile work environments o Needs clear goals for performance o Requires strong relationship between leaders and followers

Source: Lutchman, Maharaj, & Ghanem (2012)

Situational Leaders

Strengths	Weaknesses
- Develop employees to higher levels of maturity - Create a motivated worker - The complexity of the job and the maturity of the worker influence the leadership behavior when dealing with the worker - Can maintain high productivity	- Can misinterpret the maturity state of the employee leading to unintended consequences - Can lead to situations where employee quits when maturity status and leadership behaviors are poorly aligned - Can potentially discriminate against some cultures in a diverse workforce - May not effectively operate in all stages of worker development

Source: Lutchman, Maharaj, & Ghanem (2012)

Situational Leadership Model

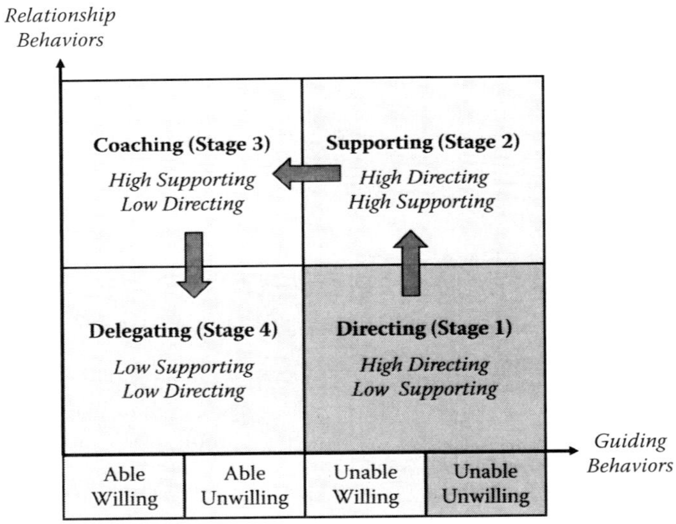

Source: Lutchman, Maharaj, & Ghanem (2012)

Stages of Situational Leadership

Leadership Behaviors	Directing / Supporting Relationship	Worker Maturity
Directing (Stage 1)	High Directing Low Supporting	**Immature – Low competence and commitment – Leadership focus on** o Telling the worker where, when, and how to do assigned work o Key requirements of structure, decision-making control, and supervision o Primarily one-way communication
Supporting (Stage 2)	High Directing High Supporting	**Immature – Growing competence; weak commitment – Leadership focus on** o Building confidence and willingness to do assigned work o Retains decision-making o Promotes two-way communications and discussions
Coaching (Stage 3)	High Supporting Low Directing	**Mature – Competent; variable commitment – Leadership focus on** o Building confidence and motivation; promote involvement o Allows day-to-day decision-making o Active listening and two-way communications and discussions
Delegating (Stage 4)	Low Supporting Low Directing	**Mature – Strong competence; strong commitment – Leadership focus on** o Promote autonomy and decision-making and empowerment o Collaborates on goal setting o Delegates responsibilities

Source: Lutchman, Maharaj, & Ghanem (2012)

Exercise

- Telling, Showing, & Engaging

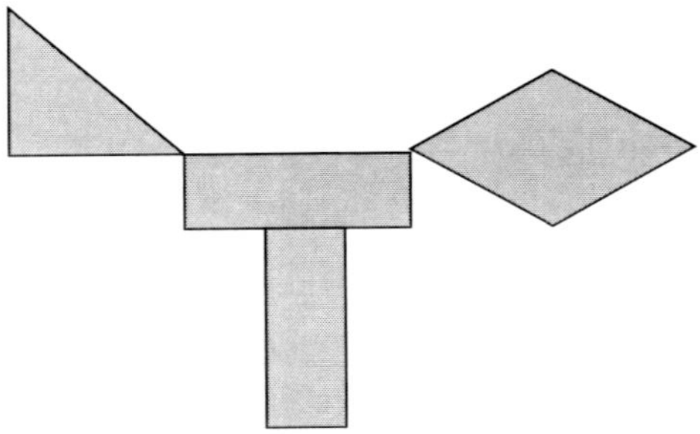

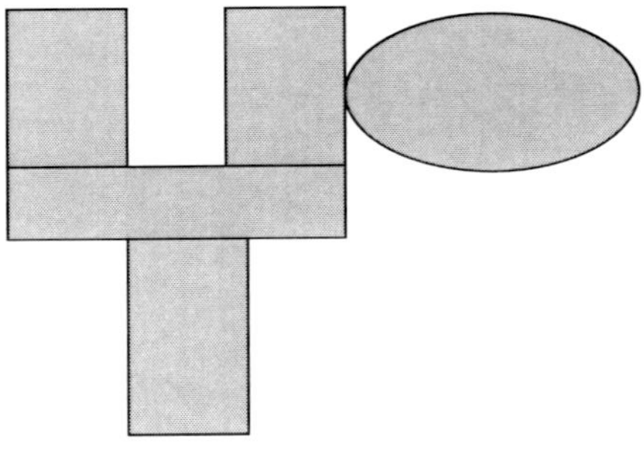

The purpose of business is to create and keep a customer.
– Peter Drucker

Transformational Leaders

Strengths

- Creates a shared vision; promotes involvement, consultation, and participation
- Encourages creativity, innovation, pro-activity, responsibility, and excellence
- Has moral authority derived from trustworthiness, competence, sense of fairness, sincerity of purpose, and personality
- Leads through periods of challenges, ambiguity, and intense competition or high growth periods
- Promotes intellectual stimulation and considers the individual capabilities of each employee
- Is willing to take risks and generate and manage change
- Leads across cultures and international borders
- Builds strong teams while focusing on macro-management
- Is charismatic and motivates workers to strong performance

Weaknesses

- Leaves a void in the organization if followers are not developed to assume role

Source: Lutchman, Maharaj, & Ghanem (2012)

Factors Influencing Motivation in the Workplace

- Organization's approach to CSR
- The personal needs of the employee – Maslow's Hierarchy of Needs
- The work environment characteristics
- The responsibilities and duties of the worker
- The level of supervision provided to the worker
- The extent of worker effort required to perform assigned tasks
- The employee's perception of organizational fairness and equity
- Career development and advancement opportunities

Source: Lutchman, Maharaj, & Ghanem (2012)

Factors Influencing Selection of Frontline Supervisor

1. Length of service with the organization
2. Amount of training the worker has been exposed to
3. Reliability of the worker in completing work
4. Attendance of the worker
5. Technical skills and competence of the worker
6. The age of the worker
7. Relationship with middle managers
8. Relationship with frontline co-workers
9. Personality
10. Availability to assume the role

Source: Lutchman, Maharaj, & Ghanem (2012)

Developing the Frontline Supervisor

1. *On-boarding*
   - Period when the worker is new to the role
   - Ideally a pre-job period exercise
   - Frontline supervisor is not yet performing supervisory work in the role
2. *Mentoring*
   - Immediate period of post on-boarding
   - Supervisor/leader is working in the role but under the guidance of a mentor
   - Mentor supports and assists the new supervisor/ leader in the role
3. *Developing in role*
   - Frontline supervisor growing in competence
   - Allowed to function alone on an ongoing basis
   - Continuous guidance from his/her leader is required to ensure success
4. *Sustainment*
   - Supervisor leader fully competent in the role
   - Continuous improvements
   - Consistent with the organizational values and behaviors

Source: Lutchman, Maharaj, & Ghanem (2012)

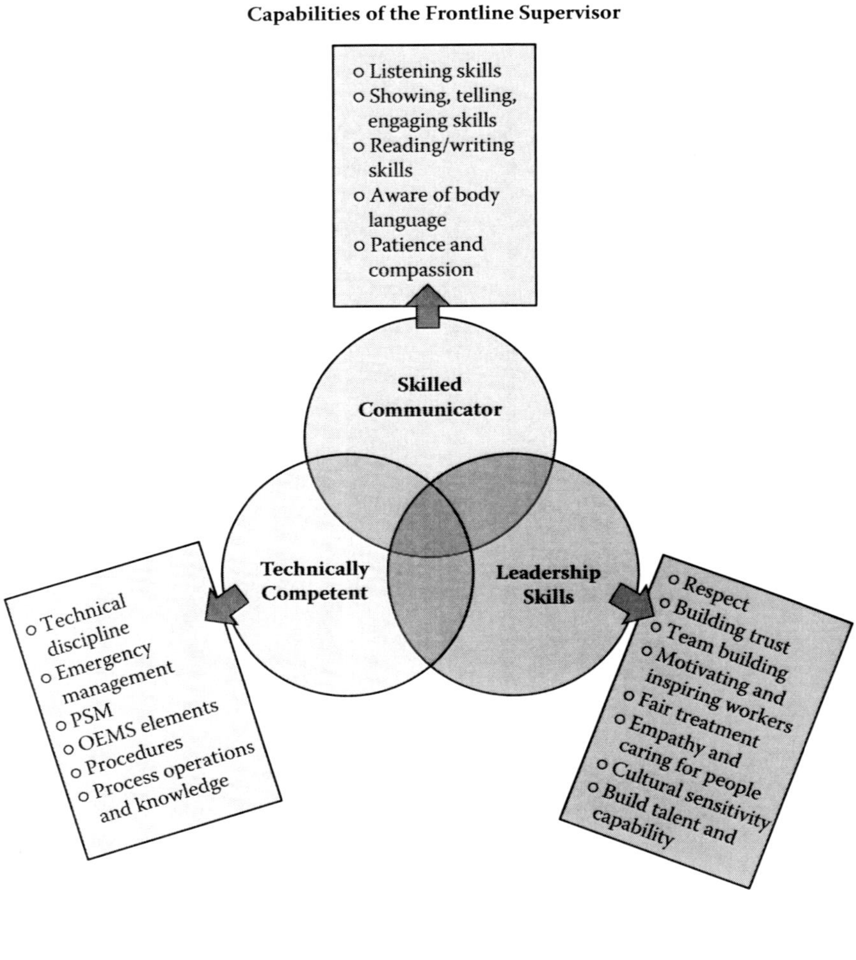
Capabilities of the Frontline Supervisor
o Listening skills
o Showing, telling, engaging skills
o Reading/writing skills
o Aware of body language
o Patience and compassion
Skilled Communicator
Technically Competent
Leadership Skills
o Technical discipline
o Emergency management
o PSM
o OEMS elements
o Procedures
o Process operations and knowledge
o Respect
o Building trust
o Team building
o Motivating and inspiring workers
o Fair treatment
o Empathy and caring for people
o Cultural sensitivity
o Build talent and capability

Leadership Traits of the Frontline Supervisor

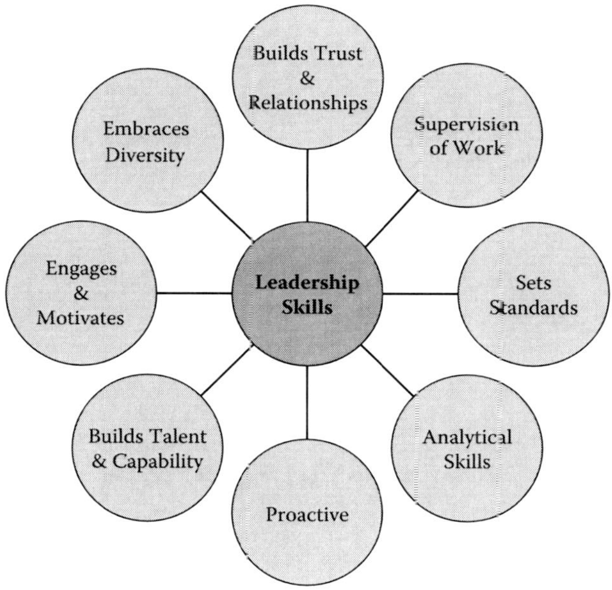

Source: Lutchman, Maharaj, & Ghanem (2012)

Essential Management Attributes for OEMS

- *Being proactive*
  - Understand the business environment (BE) needs
  - Flexibly adjustment to the changing BE
- *Having clearly defined goals*
  - Goals and targets defined
  - SMART goals that are stewarded
- *Prioritize activities*
  - Prioritization based on risk
  - Impact to business
  - Stakeholder groups
- *Create win-win outcomes*
  - Stakeholder engagement and involvement
  - Optimal solutions
- *Understand firstly then seek to be understood*
  - Communication
  - Body language
  - Listening
- *Take advantage of synergies*
  - 1 + 1 = 3
- *Personal development*
  - Continue to upgrade skills to meet business environment requirements

Situ-Transformational Leadership Model

- Combines transformational leadership behaviors and Situational Leadership Model
- Transformational leadership behaviors applied at each stage of the worker's development

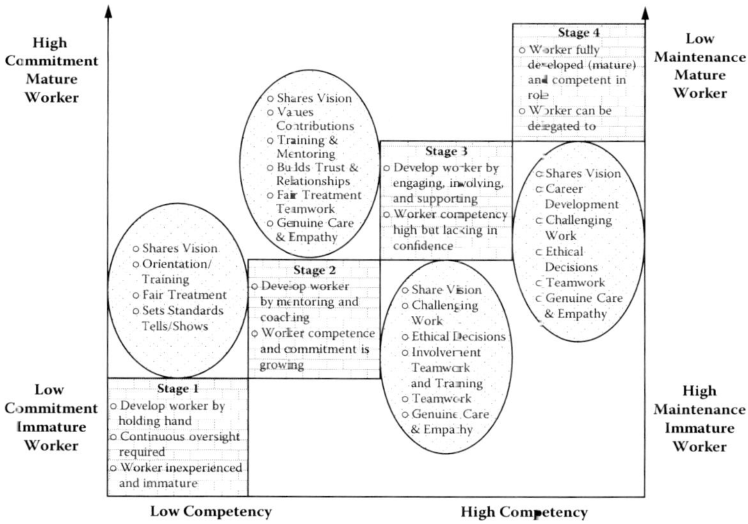

Source: Lutchman, Maharaj, & Ghanem (2012)

Corporate Values and Associated Leadership Behaviors

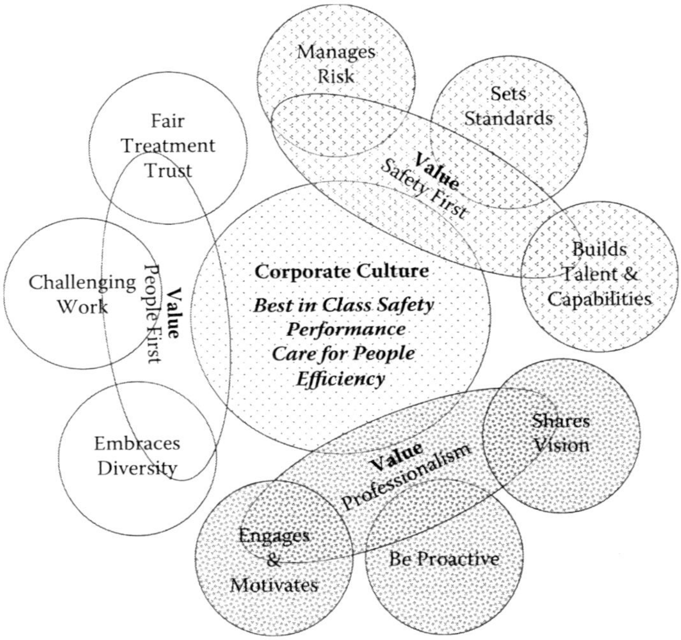

Source: Lutchman, Maharaj, & Ghanem (2012)

Key Leadership Focus for OEMS

- *Sharing the vision*
    - Senior leadership communication
    - Town hall meetings
    - Engagement sessions
- *Managing change*
    - Identifying and engaging stakeholders
    - Stakeholder impact assessment
        - Business unit
        - Business area
        - Departmental
        - Individual
- *Genuine Care & Empathy*
    - Changing work loads
    - Different ways of doing things
    - Personnel impact
- *Teamwork*
    - Senior leadership communication
    - Town hall meetings
    - Engagement and involvement sessions
- *Developing workers*
    - Directing
    - Supporting
    - Coaching
    - Delegating

Fundamental 2
Elements of a Management System

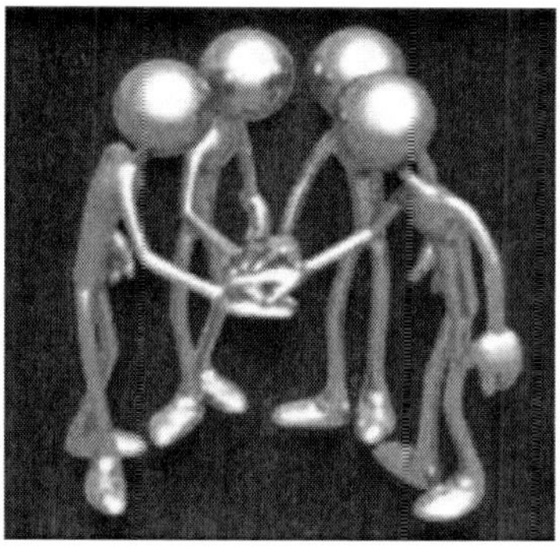

In this section you will gain an insight into the Elements of an Operationally Excellent Management System. Generally, each Element is supported by a written Policy, Standard, or Set of Guidelines and is tied to the shared vision of the organization.

Discussion Topics

- Shared Vision
- Elements of a Management System
  - Personnel Safety
  - Process Safety Management
  - Operations Management System

The future does not belong to those who are content with today ... Rather, it will belong to those who can blend vision, reason, and courage in a personal commitment.

– Peter Wagner

Start with a Shared Vision

Vision Statements

Used to depict the future picture of your company. A vision statement defines the aspirations of the organization and provides the broad river-banks for your long-term strategic planning.

A vision statement can be enterprise-wide or may be limited to a business unit or area.

A vision statement provides general corporate direction and answers the question: ***Where do we want to be in the future?***

Sample Vision Statements

DuPont

The vision of DuPont is to be the world's most dynamic science company, creating sustainable solutions essential to a better, safer and healthier life for people everywhere.

Chevron

At the heart of The Chevron Way is our vision: to be the global energy company most admired for its people, partnership, and performance.

GM

GM's vision is to be the world leader in transportation products and related services. We will earn our customers' enthusiasm through continuous improvement driven by the integrity, teamwork, and innovation of GM people.

Honda

1970: We will destroy Yamaha.

Current: To be a company that our shareholders, customers, and society wants.

Unilever

The four pillars of our vision set out the long-term direction for the company – where we want to go and how we are going to get there:

- We work to create a better future every day
- We help people feel good, look good, and get more out of life with brands and services that are good for them and good for others
- We will inspire people to take small everyday actions that can add up to a big difference for the world
- We will develop new ways of doing business with the aim of doubling the size of our company while reducing our environmental impact

We've always believed in the power of our brands to improve the quality of people's lives and in doing the right thing. As our business grows, so do our responsibilities. We recognize that global challenges such as climate change concern us all. Considering the wider impact of our actions is embedded in our values and is a fundamental part of who we are.

Strong Vision Statements	Weak Vision Statements
o Short o Focused o Clear o Memorable o Easy to understand	o Long o Unfocused o Unclear o Easy to forget o Difficult to understand

Thoughts to ponder

- Who determines the corporate vision?
- Do you know your company's vision?
- How does your business unit/area support the corporate vision?
- What is the difference between vision and mission?

Shared Vision

A shared vision

- Uplifts people's aspirations
- Harnesses and channels energy and excitement
- Provides direction and motivation

To create a shared vision

- Engage and involve people
- Start with leadership at the highest level
- Communicate ... communicate ... communicate then communicate again
- Be realistic – ensure alignment with resources capabilities (no pie in the sky)

Sample shared visions

- Journey to Zero – Suncor Energy Inc
- Beyond Zero – Jacobs

In Alice in Wonderland, *Alice gets to a fork in the road and she asks the cat:*
"Which way do I go?"
"Where are you going?" asks the cat.
"I don't know."
"Then it really doesn't make any difference," replies the cat.

Vision vs. Mission

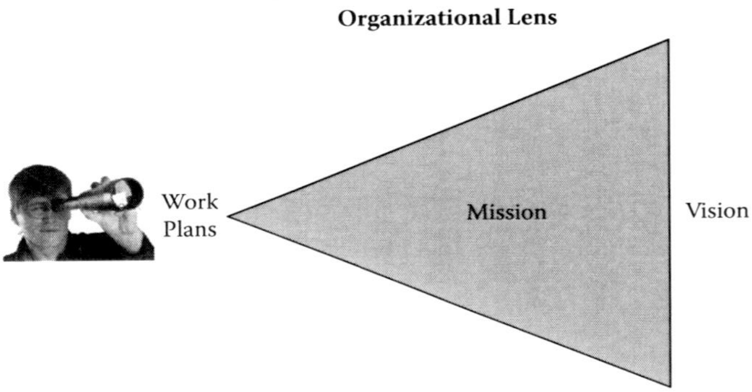

Plans	Tactical Day-to-Day Activities	Strategic & Tactical 1–3 yr Business Plan	Strategic 10–15 yr Business Plan
Actions	Clear & Specific	Clear	General
Focus	Individual	Business Unit/Area	Corporate

Elements of an Operations Management System

OMS

ELEMENTS

- Leadership, management leadership, and organizational commitment
- Risk management
- Goals, targets, and planning
- Environmental management
- Emergency preparedness and security management
- Qualification, orientation, and training
- Contractor/supplier management
- Event management and learning
- Management of engineered change and nonengineered change
- Management of personnel change
- Prestart-up safety reviews
- Operations integrity audit
- Process safety information
- Process hazard analysis
- Physical asset system integrity and reliability
- Operating procedure and safe work practices
- Stakeholder management
- Legal requirements and commitments
- Management review

Types of Management Systems

- Environmental Management System (EMS)
- Lean Integration
- Occupational Health & Safety Management System (OHSMS)
- Total Quality Management (TQM)
- Welfare Management System (WMS)

Management System Disaggregated

Management Systems	Elements
Operationally Excellent Management System (OEMS)	1. Management, leadership, and organizational commitment 2. Risk management 3. Goals, targets, and planning 4. Stakeholder management 5. Legal requirements and commitments 6. Management review
Process Safety Management (PSM)	1. Management of engineered, non-engineered, and people change 2. Prestart-up safety reviews 3. Audits & assessments 4. Process safety information 5. Process hazard analysis 6. Physical asset system integrity and reliability
Personnel Safety Management	1. Emergency preparedness and security management 2. Qualification, orientation, and training 3. Contractor/supplier management 4. Operating procedure and safe work practices 5. Event management and learning

Safety Management System = Personnel Safety Management
+
Process Safety Management

OEMS = SMS on **Steroids**

Operations Excellence = OEMS + Operations Discipline

Personal Safety Management

Definition

Personal safety management seeks to secure the health and safety of all workers during the performance of work.

- Addresses industrial hygiene needs
- Considers human factors – human workplace interface
- Injury and illnesses prevention
- Management of return to work after incidents or injuries may have occurred

Common Workplace Programs

- Safety vision and mission
  - Journey to Zero
  - Beyond Zero
  - Zero Harm
- Safety culture focus
- Personal protective equipment (PPE)
- Worker engagement and involvement in safety – ownership

Workplace Programs Focus

- Behavior observation and modification
- Housekeeping
- Training and competency assurance
- Mentoring and job shadowing

Process Safety Management (PSM)

What is PSM?

- PSM is a management process that allows leaders to proactively recognize, understand, and control process hazards, such that process-related injuries and incidents may be prevented
- PSM, when simplified, is the application of a management system and supporting tool to eliminate, prevent, and minimize the impact of process-related incidents and events
- PSM forms a critical part of an organization's Safety Management System and is designed to work in conjunction with Personnel Safety Management

OSHA (2010)

- Prevent the unintended release of hazardous chemicals that can cause harm to employees
- A PSM program must be systematic and holistic in its approach to managing process hazards and must consider:
    - The process design
    - Process technology
    - Process changes
    - Operational and maintenance activities and procedures
    - Nonroutine activities and procedures
    - Emergency preparedness plans and procedures
    - Training programs
    - Other elements that affect the process

Process Safety Management – Standards (OSHA)

- Process Safety Information (PSI)
    - Hazards of the chemicals used in the processes
    - Technology applied in the process
    - Equipment involved in the process
    - Employee involvement
- Process Hazard Analysis (PHA)
- Operating procedures
- Employee training and competency
- Contractor safety management
- Prestart-up safety review
- Mechanical integrity of equipment
    - Process defenses
    - Written procedures
    - Inspection and testing
    - Quality assurance
- Nonroutine work authorizations
- Management of change (MOC)
- Incident investigation
- Emergency preparedness planning and management
- Compliance audits
    - Planning
    - Staffing
    - Conducting the audit
    - Evaluation and corrective actions

Organizational Management System Maturity Model

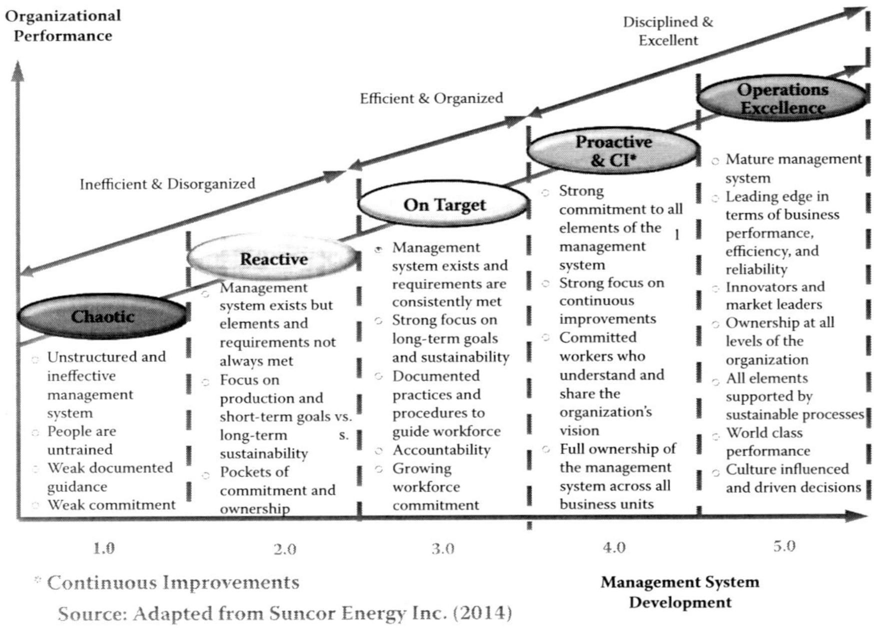

Source: Adapted from Suncor Energy Inc. (2014)

- Each element requirement must be accompanied by an implementation guideline to ensure consistent interpretation and field application
- Applicable scores are defined by the level of compliance to the element requirements
- At the very minimum, the organization must strive for ***Planned** Compliance and Development Levels*

ELEMENTS: People, Processes,& Systems, and Facilities & Technology

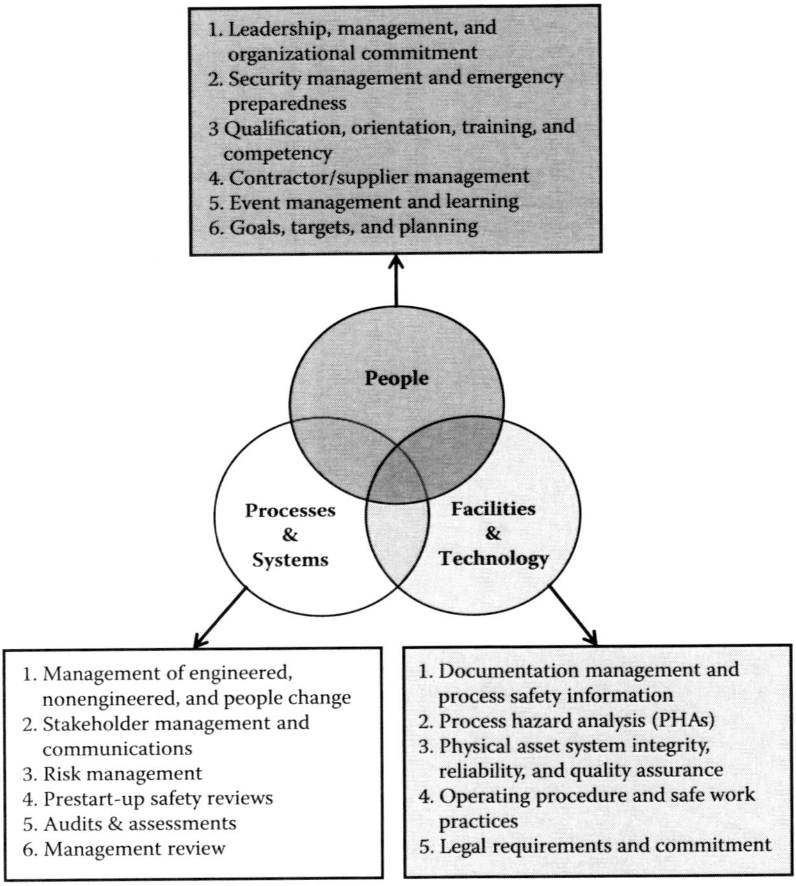

People Elements

1. Leadership, management, and organizational commitment
2. Security management and emergency preparedness
3. Qualification, orientation, and training
4. Contractor/supplier management
5. Event management and learning
6. Goals, targets, and planning

People Element

1. Leadership, Management, and Organizational Commitment

- Establishes clear expectations for followers in implementing and sustaining the management system requirements
- Assumes accountability for maintaining performance
- Drives continuous improvement to the management system and performance on an ongoing basis
- Demonstrates commitment to the management system and provides support to followers as necessary

Leadership

- Sharing the vision
- Leveraging the strategic plan
- Inspiring hearts and minds of workers
- Transformational, visible, and active behaviors
- Providing resources and support to enable performance

Management & Commitment

- Stewarding the day-to-day activities
- Executing the tactical plans and work plans
- Delivering on key performance indicators (KPIs)
- Managing risks, protecting assets and people, and maximizing production

People Element

2. Security Management and Emergency Preparedness

- Addresses planning and response related to
    - Emergencies
    - Crisis management
    - Operational and security emergencies
    - Business continuity requirements
        - Pandemic
        - Natural disasters
        - Strikes/lockouts

Structure / Training & EMS Drills

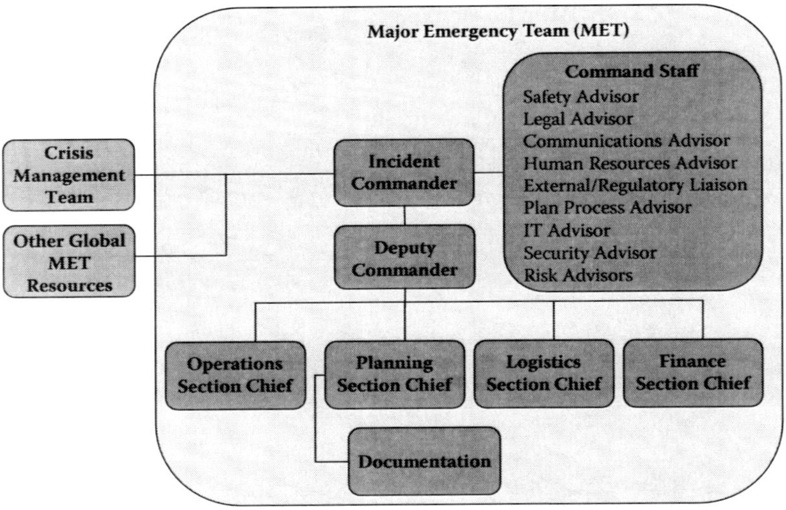

Source: Suncor Energy Inc. (2011). with approval.

People Element

3. Qualification, Orientation, and Training & Competency

- Address business needs for:
    - Identifying and implementing learning activities
    - Defining training and competency requirements to work in a particular area of the business
    - Assessing competency assurance
    - Maintaining the *collective competence* in work areas
    - Establishing the orientation requirements for new workers

Qualification

- Experience
- Skills
- Training and competency

Orientation

- Workplace hazards and risk exposures
- Varies by business areas
- Caters for contractors and employees
- Cost issues

Training & Competency

- On the job
- Mentoring and job shadowing
- Classroom/workshops/external & internal courses
- Training ≠ Competency
- Competency is a demonstration of skills, knowledge, and abilities

People Element

3. Qualification, Orientation, and Training & Competency

Collective Competence – What does it mean?

- Competence of a team or workgroup
- ≠ the summation of the individual competence of each member
- Focus on the group's collective knowledge or abilities, which exceeds that of each individual member
- Strong interdependency required among team members

New and Inexperienced Worker Orientation

- Provides a copy of the company health and safety policy to the worker
- Provides an overview of the work site/project and explains the worker's duties
- Informs the worker of *all hazards* on site and the protective measures required
- Ensures access to personal protective equipment and explains why PPE is required
- Explains the emergency response procedure, identifies the muster points and the basis for *incident* reporting
- Ensures the worker is knowledgeable on the location of emergency equipment, he/she is competent in using them, and understands the access to first aid kit and basic health response needs
- Introduces the worker to his or her supervisor and other work colleagues
- Provides a guided tour of the worksite

People Element

3. Qualification, Orientation, and Training & Competency

Training Method Effectiveness

- On the job training, mentoring, and coaching is most effective
- Involvement and engagement
- Showing in the field

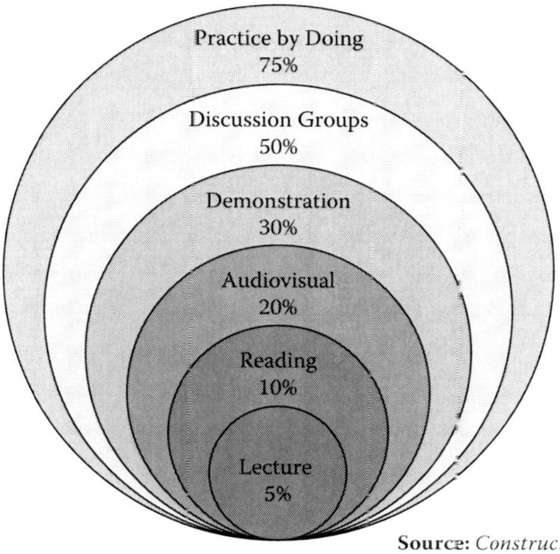

Source: *Construction Safety Magazine*

People Element

4. Contractor/Supplier Management

- Establishes requirements for the following:
    - Developing a database of prequalified contractors competent and capable of working for employer
        - Prequalified on the basis of
            - EH&S
            - Quality assurance
            - Technical competence and capabilities
            - Financial capabilities and solvency
    - Establishing and maintaining collaborative relationships with contractors/suppliers
    - Establishing and stewarding to performance management targets

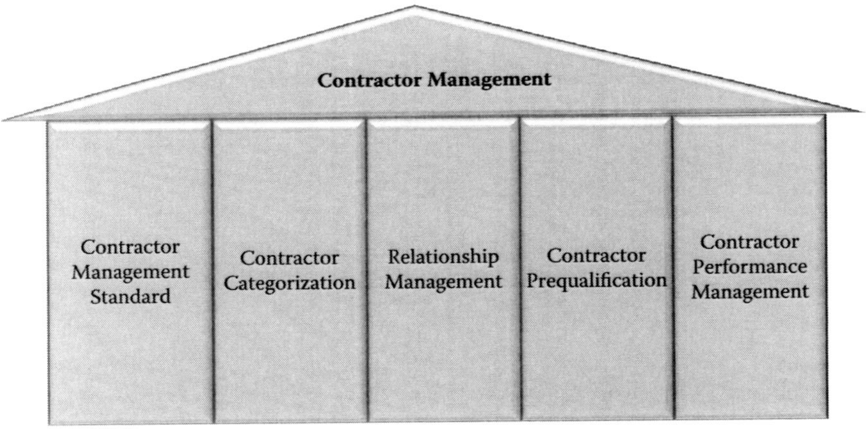

People Element

4. Contractor/Supplier Management

Stakeholder Interest Map for Each Stage of the Contracting Life Cycle

- ~80% of work performed by contractors
- Contractors perform the 3-D jobs (**D**irty, **D**ifficult, and **D**angerous)

Stakeholder Groups / Contract Life Cycle Activities	SCM	Business Unit Leaders	Frontline Supervisors & Managers Contractors/ Organization	Support Services, e.g., EH&S/ Legal/ Finance	Contractor Leadership
Contracting Strategy	Satisfaction	Scope of work	Engaged	Engaged	Considered
Contractor Prequalification***	Quality Technical Finance EH&S Scalable	Simple Reliable User friendly Credible Training	User friendly Simple Training Support	Flexible Scalable Credible	Simple User Friendly Inexpensive
Contractor Selection	Cost/Safety /Quality/ Financially stable	Reliable/ Cooperative /Easy to work with	Follows procedures Safe	Collaborative	Considered
Contractor Mobilization	Engaged	Lead	Involved	Verified	Satisfied
Performance Management	Advised	Leading Indicators	TRIF/CRIF /CDIF	Consulted	TRIF/ CRIF/ CDIF
Contractor Closeout	Share Knowledge	Feedback	Consulted	Liability	Repeat Business

Source: Lutchman, Maharaj, & Ghanem (2012)

People Element

4. Contractor/Supplier Management

Categorizing Contractors

- Owner contractor relationship changes with categorization
- Goal is to reduce risk and spend at all times
- High risk/spend requires collaborative relationships

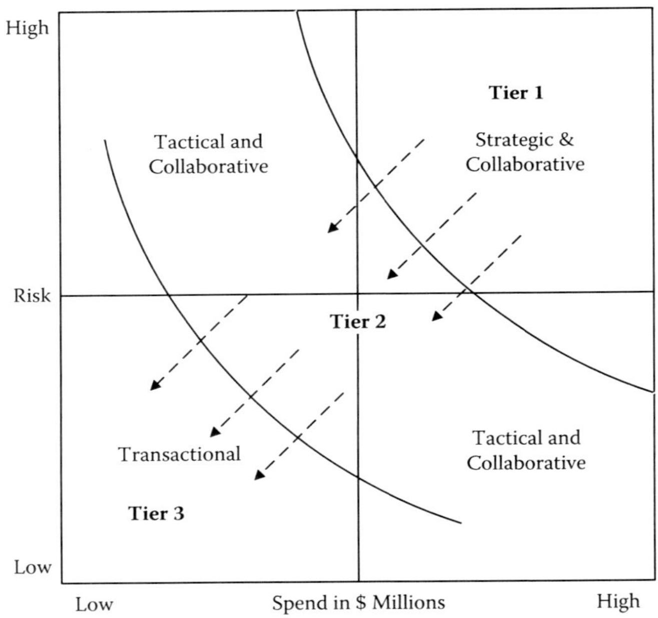

Source: Lutchman, Maharaj, & Ghanem (2012)

People Element

4. Contractor/Supplier Management

Requirements for Contractor Safety Management

- Standard
- RACI
- EH&S prequalification
- EH&S performance management

Contractor Safety Standard

- Defines purpose and scope
- Defines roles and responsibilities
- Contractor safety management requirements

RACI

- Removes ambiguity
- Clearly defined responsibilities, accountabilities, consultation, and information needs

EH&S Prequalification

- Prequalification questionnaire
- Selecting your prequalification service provider
- Activating your prequalification service provider

EH&S Performance Management

- Supporting tools and processes
- KPI stewardship
- Auditing of contractors, processes, and performance

Source: Lutchman, Maharaj, & Ghanem (2012)

People Element

4. **Contractor/Supplier Management**

Contractor Safety Audits Highly Beneficial

- Contractors want to do a good job and protect their workforce
- Contractors want to work collaboratively with owners to identify and close gaps in their safety management systems
- Collaborative audits improve EH&S performance

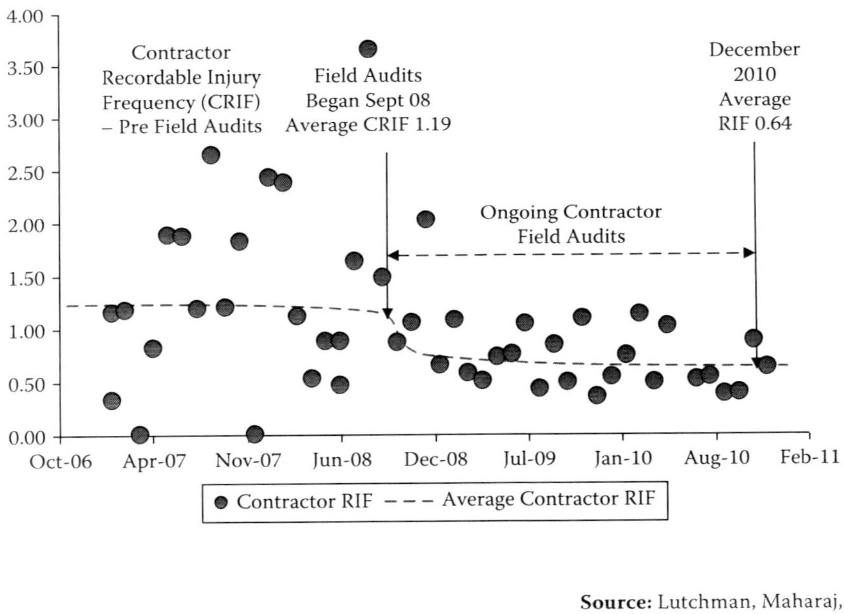

Source: Lutchman, Maharaj, & Ghanem (2012)

People Element

5. Event Management and Learning

- Formal processes required for the reporting, investigation, and subsequent management of incidents and hazards
    - Incidents and hazards reporting
    - Incidents and hazards risk assessments
    - Incident investigations and root cause analysis
        - Tap root
        - 5 Whys
        - DNV
        - Reason
    - Corrective actions management
        - Accountable/responsible
        - Priority
        - Approval and resourcing
    - Knowledge generated and Learning from Events (LFE) processes
        - Trend analysis and data management and stewardship
        - Leading vs. lagging indicator focus

People Element

5. Event Management and Learning

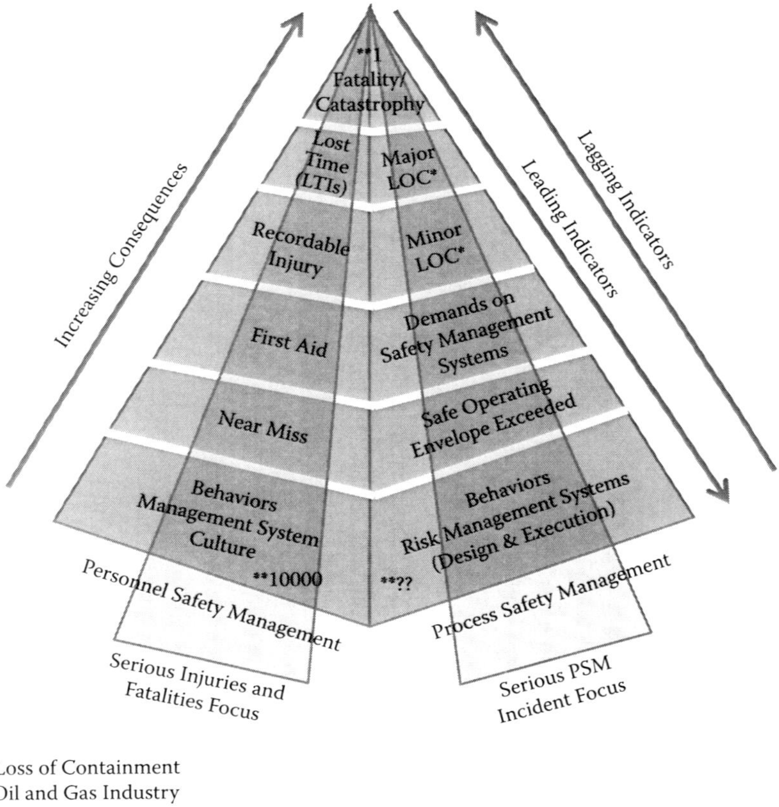

*Loss of Containment
**Oil and Gas Industry

People Element

6. Goals, Targets, and Planning

- Establishes requirements for setting of goals and targets to develop business plans and to assist in the understanding of expected contributions, priorities, and deliverables

Setting Goals and Targets

- Involvement of key stakeholders
- SMART – **S**pecific, **M**easurable, **A**chievable, **R**ealistic & **T**ime-bound
- Aligns with the strategic vision

Business Planning

- Strategic Plan
  - 5–10 horizon
  - Depends on the business environment indicators
  - Focus on projected Capex/Opex/Cash flow/IRR/NPV/Manpower Loading
- Tactical/Annual Plan
  - Focus on monthly/weekly/daily stewardship
  - KPIs
  - 3+9, 6+6, 9+3 adjustments
  - Cascades into each worker's work plan
- Day-to-Day Plan
  - Managing the day-to-day issues to achieve tactical plan
  - Delivering on the employee work plan

Facilities & Technology Elements

1. Documentation management and Process Safety Information (PSI)
2. Process hazard analysis (PHAs)
3. Physical asset system integrity, reliability, and quality assurance
4. Operating procedure and safe work practices
5. Legal requirements and commitment

Facilities & Technology Element

1. Process Safety Information (PSI) and documentation management

- PSI allows the organization to manage information related to the sustained operations of its business. Business units and operating areas are required to have in place processes for identification, control and management of all data, documents, and information deemed to be critical for sustained operations of the business
    - **Hazards of materials**
        - Lists chemical and physical data
        - Lists raw materials, intermediates, waste products, finished products
    - **Process design basis**
        - Describes process chemistry
        - Includes process steps and limits
        - Designates maximum intended inventory or hazardous materials
        - Includes consequences of deviation from established limits
    - **Equipment/asset design basis**
        - Designates process safety-critical equipment
        - Describes vital equipment design data
        - Describes codes and standards used in fabricating the equipment

Key Requirement

- Formal processes for managing documentation and data
- Easily accessible
- Approved, current, and updated
- Reviewed and updated on a predetermined frequency
- Controlled access and security

Facilities & Technology Element

1. Process Safety Information (PSI) and documentation management

- Managing the document life cycle from creation through retirement is a very complex and difficult process. Continuous attention and perseverance is required to get it right.

Challenges in Document Management System

- Storage of document at each stage of its life cycle
- Control and access to documents at each stage of its life cycle
- Managing documents across a diverse organization with geographically dispersed team members who contribute to the documents' creation, review, approval, disbursement, and storage
- Document policies regarding access to sensitive information, retention, retirement, and distribution
- How to handle documents as corporate records, which must be retained according to legal requirements and corporate guidelines

Document Life Cycle Management

Creation & Approval

- *Authors* – Personswho create the document
- *Reviewers* – Subject matterexperts who contribute to the accuracy and alignment with intended purpose
- *Approvers* – Persons who approvethe document for usage

Issue & Use, Change, Storage & Distribution

- *Issue & Use* – Onlyafter approval
- *Change* – Followthe MOC process to ensure a managed change process
- *Storage & Distribution* – Accessiblecommon server with user access

Security, Maintenance, and Retirement & Disposal

- *Security* – User access reflected in security considerations on a per user basis
- *Maintenance* – Updates ad perpredetermined frequency
- *Retirement & Disposal* – Retirementas per regulations and policies and disposal consistent with sensitivity and security need

Facilities & Technology Element

1. Process Safety Information (PSI) and documentation management

Leveraging Technology

- Use simple computerized processes as much as possible
- Control and manage access and distribution and read-write access
- Maintain version control during creation
- If not for public viewing and sharing, maintain high security
- Consistent nomenclature and search capabilities

o Manual o Labor Intensive
o Paper o Centralized

o Computerized o Easily accessible
o Paperless o Distributed

Types of Documentation to Be Managed

- Process flow diagrams (PFDs), Piping and Instrumentation Diagrams (P&IDs)
- Engineering specifications
- Incident investigation reports and best practices
- Maintenance and Standard Operating Procedures (SOPs)
- Training and competency assurance records
- Chemical processes & MSDS data

Facilities & Technology Element

2. Process Hazard Analysis (PHA)

- PHA is an effective technique to systematically identify, evaluate, and develop methods to control hazards associated with hazardous processes at various times and stages in the life cycle of a process. These hazards generally represent the potential for fires, explosions, and/or the release of toxic materials.

PHAs Generally Applicable to

- Newly constructed hazardous processes – baseline process hazard analysis
- Existing assets that do not have a baseline process hazard analysis
- Existing assets – cyclic process hazard analysis or revalidation
- Assets being mothballed
- Assets being dismantled
- Drilling and well-work operations

PHA Methodology

- "What If"
- Checklist
- "What If"/Checklist
- Hazard and Operability Analysis (HAZOP)
- Failure Mode and Effect Analysis (FMEA)
- Fault Tree Analysis (FTA)
- Existing assets – cyclic process hazard analysis or revalidation

Facilities & Technology Element

2. Process Hazard Analysis (PHA)

PHAs – Who Does It?

- Team with expertise in engineering and process operations
- Includes personnel with experience and knowledge specific to the process being evaluated
- Includes personnel with experience in the PHA methodology being used

PHAs Must Address the Following:

1. All hazards associated with the process
2. All prior incidents with likely potential for catastrophic consequences
3. Consequences of failure of engineering and administrative controls, especially those affecting employees
4. Engineering and administrative controls applicable to each hazard and the interrelationship among hazards and controls
5. Facility location and siting
6. Human factors considerations
7. Action management to promptly resolve all high-priority PHA findings and recommendations

Factors Influencing Methodology Selection

- The amount of existing knowledge about the process
    - Has the process been operated for an extensive period with little or no innovation and considerable experience has been generated?
    - Is it a new process?
    - Has there been many changes due to innovation?
    - Is the technology still evolving?
- The size and complexity of the process

Facilities & Technology Element

2. **Process Hazard Analysis (PHA)**

Steps in the PHA Process

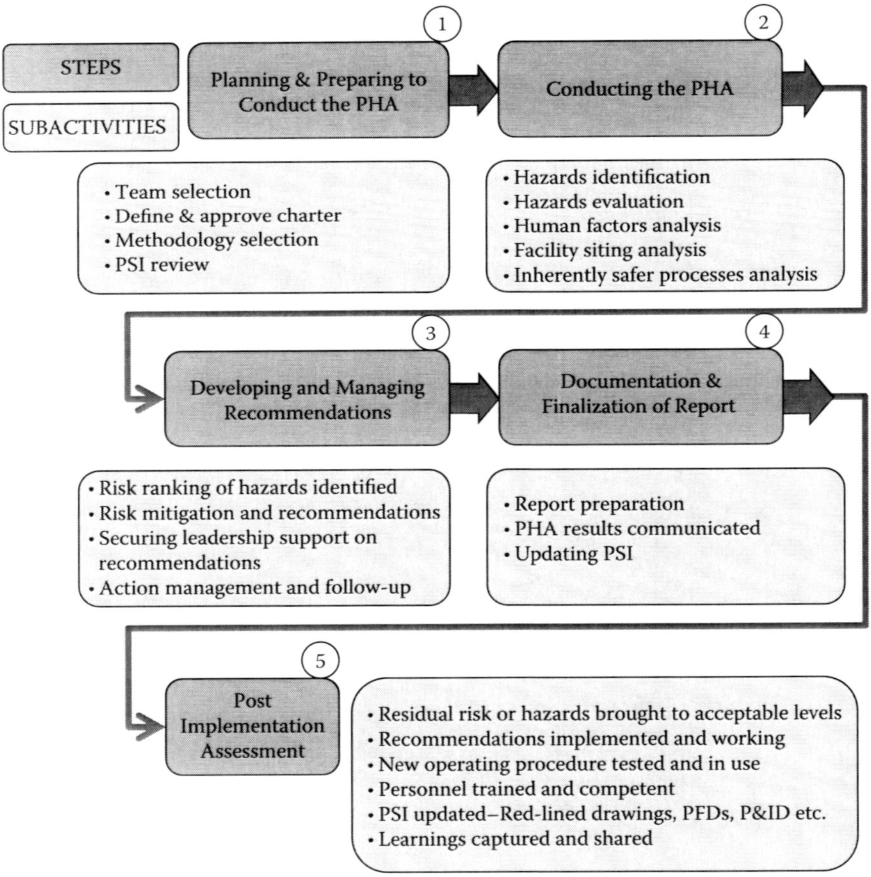

Facilities & Technology Element

2. Process Hazard Analysis (PHA)

Methodologies Summarized

"What If"

1. Experienced team brainstorming on "What if…?" scenarios/questions. Focus on the following:
   - ➢ Equipment failure: e.g., What if a pump fails?
   - ➢ Human errors: e.g., What if an operator forgets to close a valve?
   - ➢ External factors: e.g., What if there is a lightning strike?

1. Each scenario/question represents a potential failure or maloperation of the facility
2. Responses of the process and personnel assessed to determine the extent of the hazard created
3. Existing safeguards weighed against the probability and severity of the scenario to determine whether modifications to the system should be recommended

Steps

1. Divide process/system into logical subsystems
2. Develop list of questions for each subsystem
3. For each question, identify hazards, consequences, severity, likelihood, and recommendations
4. Repeat Steps 2–3 until complete

Facilities & Technology Element

2. Process Hazard Analysis (PHA)

Methodologies Summarized

Checklist

- Applies a prepared, detailed list of questions about the design and operation of the facility
  - Questions are usually answered Yes or No
- Used to identify common hazards through compliance with established practices and standards

Question Focus

1. Causes of incidents
   - Hardware – Process equipment – Do we have the right metallurgy?
   - Human error: e.g., Is equipment identified properly?
   - External events – Weather, etc. (e.g., Is the system designed to withstand -30°C?)

2. Facility Functions
   - Alarms, construction materials, control systems, documentation and training, instrumentation, piping, pumps, vessels, etc.
   - Questions such as, Are there alarm tone differences? Is the vessel internal corrosion measurable online? Are sources of ignition controlled? Is pressure relief provided? Is there sufficient venting capacity?

Facilities & Technology Element

2. Process Hazard Analysis (PHA)

Methodologies Summarized

"What If"/Checklist

- A hybrid of the "What if" and Checklist methodologies
- Integrates brainstorming in a structured framework

Question Focus

1. Causes of incidents
   - Hardware – Process equipment – Do we have the right metallurgy?
   - Human error: e.g., Is equipment identified properly?
   - External events – Weather, etc., e.g., (Is the system designed to withstand -30°C?)
2. Facility Functions
   - Alarms, construction materials, control systems, documentation and training, instrumentation, piping, pumps, vessels, etc.
   - Questions such as, Are there alarm tone differences? Is the vessel internal corrosion measurable online? Are sources of ignition controlled? Is pressure relief provided? Is there sufficient venting capacity?
3. Brainstorming questions
   - May generate new questions outside the Checklist questions but very relevant to the hazards of the process

Facilities & Technology Element

2. **Process Hazard Analysis (PHA)**

Methodologies Summarized

Hazard and Operability Analysis (HAZOPs)

1. Break operating system and process equipment into smaller operating sections such as a vessel or pump and associated piping and instrumentation
2. Identify all hazards (safety, health, environmental)
3. Identify potential problems that may prevent the efficient operation of the system or asset under review
   - Identify deviations from normal operations and potential hazards introduced from such deviations as well as the potential consequences
4. Identify potential failures that may lead to deviation
5. Ensure failure detection and mitigation systems are adequate
6. Identify recommendations
7. Repeat for all sections for the process until entire process is completed
8. Documentation

Types of Deviations in (HAZOP) Studies

- Pressure exceeds design envelope
- Pressure beneath design envelope
- Temperature exceeds design
- Temperature falls beneath design
- High and low flow scenarios
- Equipment failures, deterioration, or design

Facilities & Technology Element

2. Process Hazard Analysis (PHA)

Methodologies Summarized

FMEA – Failure Modes Effects Analysis

1. Manual analysis to determine the consequences of component, module, or subsystem failures
2. Bottom-up analysis
3. Consists of a spreadsheet where each failure mode, possible causes, probability of occurrence, consequences, and proposed safeguards are noted

Types of Failures for Consideration	
o Ruptured piping, gaskets	o Loss of function
o Cracked piping, vessels	o Over pressure
o Leaking flanges, valves	o Vacuum creation
o Plugged lines, drains	o Temperature exceedances
o Failure to open	o Low temperature
o Failure to close	o Overfilling
o Pump failure to stop	o Bypass operations
o Pump failure to start	o Loss of instrumentation
o Spurious starts & stops	o Instrumentation bypassed

Facilities & Technology Element

2. **Process Hazard Analysis (PHA)**

Methodologies Summarized

FMEA – Failure Modes Effects Analysis

1. Sample FMEA – Exchanger Failure

Failure Mode	• Tube ruptures
Cause of Failure	• Corrosion from shell side fluids
Leading Operating Conditions	• Corrosion inhibitor pump failure
Predicted Frequency	• Occurred 2 times in previous 10 years
Failure Impact and Consequences	• Catastrophic failure with potential fatality and major fire

- Risk rank each hazard (frequency × impact or consequence)
- Identify safeguards for high and unacceptable risk items

Facilities & Technology Element

2. Process Hazard Analysis (PHA)

Methodologies Summarized

Fault Tree Analysis

1. Visual presentation that starts with a hazardous event and works backward to identify the causes of all high-priority potential events
2. Very much like a root cause analysis based on hypothetical scenarios
3. All faults identified and recommendations made to prevent fault occurrence

Existing Assets — Cyclic PHA or Revalidation

- Revalidations must determine whether the current PHAs continue to be accurate or require updating
- Revalidation is particularly important if the current PHA did not incorporate all of the process hazard analysis elements, such as
    - Consequence Analysis
    - Facility Citing Analysis
    - Human Factors
    - Inherently Safer Processes
    - Interlock Evaluation
- Revalidations must be conducted by the revalidation team and incorporated into the process hazard analysis documentation

Facilities & Technology Element

3. Physical Assets, System Integrity, Reliability, and Quality Assurance

Asset & Systems Integrity and Reliability

Assets and systems integrity and reliability are critical to the sustainable continuous operations of the organization's assets. What this means is that the organization must have systems and processes in place to ensure the following:

- Mechanical integrity
- Inspections, testing, and monitoring
- Quality control and assurance
- Preventive maintenance

Mechanical Integrity

Mechanical integrity (MI) focuses on *proactively maintaining* and *continuously improving the integrity* of systems containing hazardous substances throughout the life of the facility, from the initial installation through dismantlement. It addresses operating requirements such as the following:

- Maintenance and reliability (M&R) procedures
- Training and performance of maintenance and reliability personnel
- Quality control (QC) procedures and materials and service providers
- Equipment testing and inspections, including predictive and preventive maintenance
- Repairs and changes
- Reliability engineering

Facilities & Technology Element

3. Physical Assets, System Integrity, Reliability, and Quality Assurance

Asset & Systems Integrity and Reliability

Inspections, Testing, and Monitoring

Predictive and preventive maintenance programs shall be established for PSM-critical equipment consisting of inspections and tests to detect impending or minor failures and procedures to mitigate their potential to develop into more serious failures.

Essential Requirements

- Equipment list for testing and inspections
- Documentation of test objectives and methods
    - ➢ Meets recognized and generally accepted good engineering practice
    - ➢ Codes and standards shall provide guidance in determining good engineering practices
    - ➢ Areas shall document inspection and testing procedures with appropriate references to applicable codes, standards, and vendors' recommendations that were used as a basis for good engineering practices
- Inspection frequencies
    - ➢ Consistent with
        - Recognized and generally accepted good engineering practices
        - Prior operating experience
        - Risk-based methodology where applicable
    - ➢ System to track equipment tests and inspections scheduled and completed
    - ➢ Written approval required to temporarily or permanently extend or defer the scheduled test or inspection target date involving process safety-critical equipment and components

Facilities & Technology Element

3. Physical Assets, System Integrity, Reliability, and Quality Assurance

Asset & Systems Integrity and Reliability

Inspections, Testing, and Monitoring (Continued)

Predictive and preventive maintenance programs shall be established for PSM-critical equipment consisting of inspections and tests to detect impending or minor failures and procedures to mitigate their potential to develop into more serious failures.

Essential Requirements

- Acceptable performance limits
  - Consistent with limits established in the PSI database
  - Deficiencies identified as outside of acceptable limits:
    - Corrected prior to further use or operation
    - Corrected in a safe and timely manner
    - Risk mitigation plans and approvals are completed and documented to help ensure interim safe operation
  - Where required, to ensure safe operation fitness for service calculations shall be performed
- Exception lists shall be issued for corrective action and follow-up for project activities
- The inspection and test records shall be kept for the life of the equipment

Facilities & Technology Element

3. Physical Assets, System Integrity, Reliability, and Quality Assurance

Asset & Systems Integrity and Reliability

Inspections, Testing, and Monitoring (Continued)

Nondestructive Testing (NDT) Methodologies

- Radiography
    - Use of X-rays to detect inconsistencies in welding joints and welds
    - Area to be roped off and access prevented during X-ray work
- Magnetic Particles
    - Used in identifying surface and subsurface discontinuities in ferroelectric materials and alloys
    - Surface/subsurface discontinuity allows magnetic flux leaks
    - Particles build up at the area of leakage to produce an indication
    - Indication examined in greater details for corrective actions
- Ultrasonic
    - Generally used to determine the thickness of the test object
    - Used in monitoring piping corrosion

Facilities & Technology Element

3. Physical Assets, System Integrity, Reliability, and Quality Assurance

Asset & Systems Integrity and Reliability

Quality Assurance/Control

QA/QC procedures are required for each business unit and area operating hazardous processes to ensure that

- Suppliers and sub-suppliers for maintenance materials, spare parts, equipment, and services are evaluated to maintain integrity and reliability
  - Maintenance materials, spare parts, and equipment meet design specifications as delivered
  - Substitutions considered replacement-in-kind shall be subject to engineering review and/or approval
  - Nonstandard or substandard materials are not placed in process safety-critical service
- Identification and management of materials to be removed from process safety-critical service is done in a timely manner
- Maintenance and reliability services that can affect the ability of equipment to operate safely and reliably. E.g.:
  - Nondestructive testing and inspection services
  - Chemical and high-pressure water cleaning services
  - Leak repair services
  - Fire protection system testing services
  - Traceability of materials and equipment

QA/QC Procedures

- Maintained and kept current and approved at all times
- Accessible by all personnel including contract maintenance workers

Facilities & Technology Element

4. Operating Procedures and Safe Work Permits

Standard Operating Procedures (SOPs)

- SOPs are designed to guide operating personnel in performing work safely
- SOPs provide a sequential set of steps to operating personnel such that, when followed properly, a worker who is not familiar with the equipment or process will be able to safely and effectively perform an assigned task
- Where new equipment and machinery are involved in the process, vendors may provide operating manuals for these equipment and machinery

A SOP Will Typically Include Following Details:

1. Nomenclature – a reference number and title for identifying and searching for the procedure
2. Date a procedure was written
3. Author of the procedure
4. Person(s) who reviewed the procedure
5. Revision number of the procedure
6. Page numbers of the procedure – normally numbered in the format "page x of y"
7. Hazards identified and mitigations required
    - Properties of and hazards presented by the materials used in the process
    - Controls to prevent exposure, including engineering controls, administrative controls, and personal protective equipment
    - Control measures to be taken if loss of containment occurs
    - Control measures to be taken if personnel exposure occurs
    - Any special hazards (e.g., reactivity, thermal, or stored energy or pressures)

Facilities & Technology Element

4. Operating Procedures (SOP) and Safe Work Permits

SOP Details (Continued)

8. Address process steps for each of the following phases:
   - Normal start-up (including PSSR)
   - Start-ups following a turnaround or an emergency shutdown (including PSSR)
   - Emergency operations
   - Normal operations
   - Planned shutdowns
   - Emergency shutdowns (including the conditions under which emergency shutdowns are indicated and the assignment of emergency shutdowns to qualified operators to ensure that they are executed in a safe and timely manner)
   - Temporary and abnormal operations as appropriate
   - Identify those raw materials and other substances critical to the safety of the process
   - Develop quality control procedures to ensure material specifications have been met
   - Make available applicable operating limits, conditions, consequences of deviation, and corrective steps through the operating procedure
   - Specify inventory limits
   - Describe safety systems and their function (e.g., isolation valves, emergency dump valves, scrubbers, flares)
   - Describe instrument controls, including alarm and interlock set points
9. Be readily accessible to personnel who work in or maintain the process
10. Updated and approved prior to implementing any changes to chemicals, technology, or assets
11. PPE required for performing the work

Facilities & Technology Element

4. Operating Procedures and Safe Work Practices

Safe Work Permits ... What Are They?

Safe work permits and practices are required to ensure the safe conduct of nonroutine, but repetitive, operating, maintenance and construction activities.

The following activities shall be covered as a minimum:

1. Work permit and authorization
2. Entrance to and exit from a hazardous process facility by maintenance, contract, laboratory, and other support personnel
3. Breaking the integrity of process equipment and piping
4. Energy isolation – lockout/tag-out of hazardous energy sources
5. Control of ignition sources (i.e., hot work permit)
6. Entry into confined spaces
7. Movement of heavy equipment relative to equipment containing hazardous materials
8. Integrity check (e.g., pressure test) of process equipment at time of turnover from maintenance and reliability personnel, but prior to acceptance
9. Bypassing of a safety interlock or alarm
10. Continued operation with an activated safety alarm
11. Temporary service-to-process flexible hoses for purging or flushing, unplugging, or similar tasks

Facilities & Technology Element

4. **Operating Procedures and Safe Work Practices**

Searching for Answers

Important Information

Store unopened film below 24°C (75 °F). Do not freeze film. Use above 13 °C (55 °F), place developing picture in warm pocket. Warranty: Polaroid will replace film defective in manufacture, packaging, or labeling. Warranty: Does not apply to outdated film and excludes all consequential damages except in jurisdictions not allowing such exclusions or limitations. Caution: This film uses a small amount of caustic paste. If any paste appears, avoid contact with skin, eyes, and mouth and keep away from children. If you get some paste on your skin wipe off immediately and wash with water to avoid an alkali burn. If eye or mouth contact occurs, quickly wash the area with plenty of water and see a doctor. Do not cut or take apart the picture or battery. Do not burn the battery or allow metal to touch its terminals.

Search 144 Words

Source: CAT-*I* Training 2013

Facilities & Technology Element

4. Operating Procedures and Safe Work Practices

Navigating to Answers

Title: Safety and Handling Guidelines

Storing	Always store your film as follows: • Away from children and pets • In a dry place between 2–24°C
Developing	Pictures develop automatically Environmental temperature should be between 13–24°C Note: Develop in a warm pocket on a cold day
Caution	Please use caution because: • Corrosive: The film chemicals contain caustic • Electrical current: The film pack contains a battery
First Aid	First aid tips if a person comes in contact with developing chemicals in the film: **Chemical on Skin** 1. Wipe off 2. Rinse with water **Chemical in Eye or Mouth** 1. Wipe off 2. Rinse with water 3. Call a doctor
Warranty Policy	The warranty policy for this film is **We Replace the Film if** 1. Defective manufacture 2. Defective packaging 3. Incorrect labeling **We do not Replace the Film if** 1. Outdated 2. Subsequently damaged
Disposal	Dispose of this film pack: • Like batteries • Do not incinerate **Search 144 Words**

Source: CAT-*I* Training 2013

Facilities & Technology Element

4. **Operating Procedures and Safe Work Practices**

Mapping User Questions to Answers

Title: Safety and Handling Guidelines

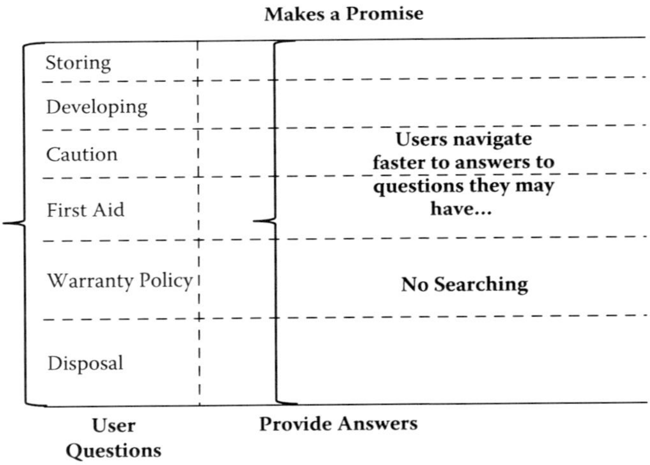

Source: CAT-*I* Training 2013

Facilities & Technology Element

4. Operating Procedures and Safe Work Practices

Cognitive Linking of Questions to Answers

Title: Safety and Handling Guidelines

User Questions	Provide Answers
Storing	Always **store** your film as follows:
Developing	Pictures **develop** automatically
Caution	Please use **caution** because:
First Aid	**First aid** tips if a person comes in contact with developing chemicals in the film
Warranty Policy	The **warranty policy** for this film is:
Disposal	**Dispose** of this film pack:

Source: CAT-*I* Training 2013

Facilities & Technology Element

4. Operating Procedures and Safe Work Practices

Cognitive Linking of Questions to Answers

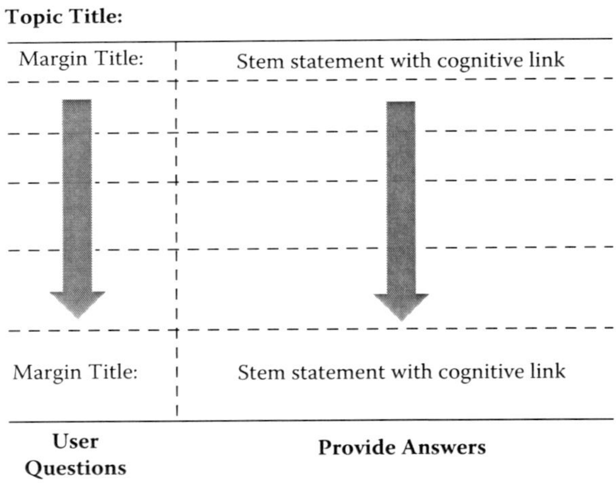

Source: CAT-*I* Training 2013

Facilities & Technology Element

4. Operating Procedures and Safe Work Permits

Types of Safe Work Permits

The following types of Safe Work Permits are necessary to permit work:

1. Cold work permits
2. Hot work permits
3. Confined space entry permits
4. Vehicle entry permits
5. Ground disturbance permits

Safe Work Permits ... They Cover (but are not limited to) the Following:

Before work can be started, the issuing and performing authorities must visit the work site and review all hazards and mitigation required before the permit is signed and approved by both parties. Conditions of the work permit must be reviewed by all members of the work crew before start of work – *signoff on the permit required by all members*

1. Work description
2. Hazards associated with the work to be completed and controls activated
    - Gas testing and monitoring
    - Energy isolation
    - Working from heights and fall arrest protection
    - PPE required
3. Issuing and performing authorities
    - Signatures of all team members of the performing authorities indicating that hazards have been reviewed, understood, and mitigated
4. Start and finish times of the work

Facilities & Technology Element

4. Operating Procedures and Safe Work Permits

Permits Coordination

Safe work permit coordination is critical and essential when multiple stakeholders inherit the worksite at the same time. Permit coordination is essential in

- Managing the workflow
- Keeping everyone safe

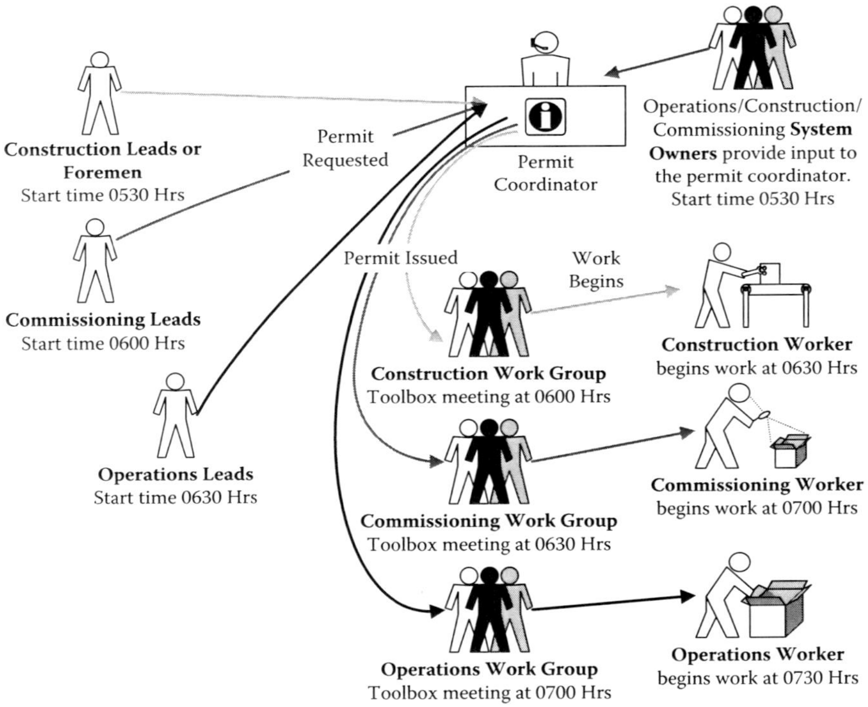

Source: Lutchman (2010)

Facilities & Technology Element

5. Legal Requirements and Commitments

Meeting and commitment legal requirements is of fundamental significance to the all organizations. This means that applicable legal requirement and commitments are identified, interpreted, and translated for action and relevance.

The Legal Compliance Process

Identification and Recording of Legal Requirements

- Process for identifying requirements
- Process for monitoring emerging requirements

Compliance

- Consistent interpretation and application
- Trained and competent personnel to drive compliance and commitment and transparency

Demonstrated Compliance with Legal Requirements

- Documentation and records
- Not only appear to be seen as compliant but be compliant with the law

Evaluation of Compliance

- Audits and assessments review
- Cold eyes review for compliance

Management Review

- Gap closure stewardship
- Resources commitment

Facilities & Technology Element

5. Legal Requirements and Commitments

Legal Commitments require the following:

- Awareness of legal requirements
- Understanding of legal requirements
- Proper interpretation of legal requirements
- Communication of legal requirements across the organization
- Compliance

Hierarchy of Documentation and Processes for Compliance

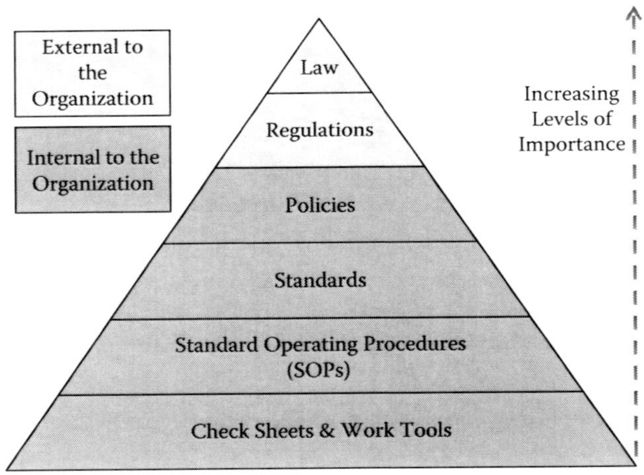

Source: Lutchman, Maharaj, & Ghanem (2012)

Processes & Systems Elements

1. Management of engineered, nonengineered, and people change
2. Stakeholder management and communications
3. Risk management
4. Prestart-up safety reviews
5. Audits & assessments
6. Management review

Processes & Systems Element

1. Management of Engineered, Nonengineered, and People Change (MOC)

Types of Changes

- ➢ Administrative
- ➢ Business Procedural
- ➢ Engineered
- ➢ Nonengineered
- ➢ Organizational
- ➢ Emergency

Change Priorities

- ➢ 1st Priority – When people, the environment, and assets are exposed to *unacceptable risks* in the order identified above
- ➢ 2nd Priority – When an emergency change is required
- ➢ 3rd Priority – All other changes

MOC Requirements

- ➢ MOC Form
- ➢ MOC team – SMEs and leadership
- ➢ Approval process and team

MOC Exemptions

- ➢ Replacement in kind
- ➢ Change in nonhazardous operations
- ➢ Procedural changes to comply with regulations

Processes & Systems Element

1. Management of Engineered, Nonengineered, and People Change (MOC)

Steps in the MOC Process

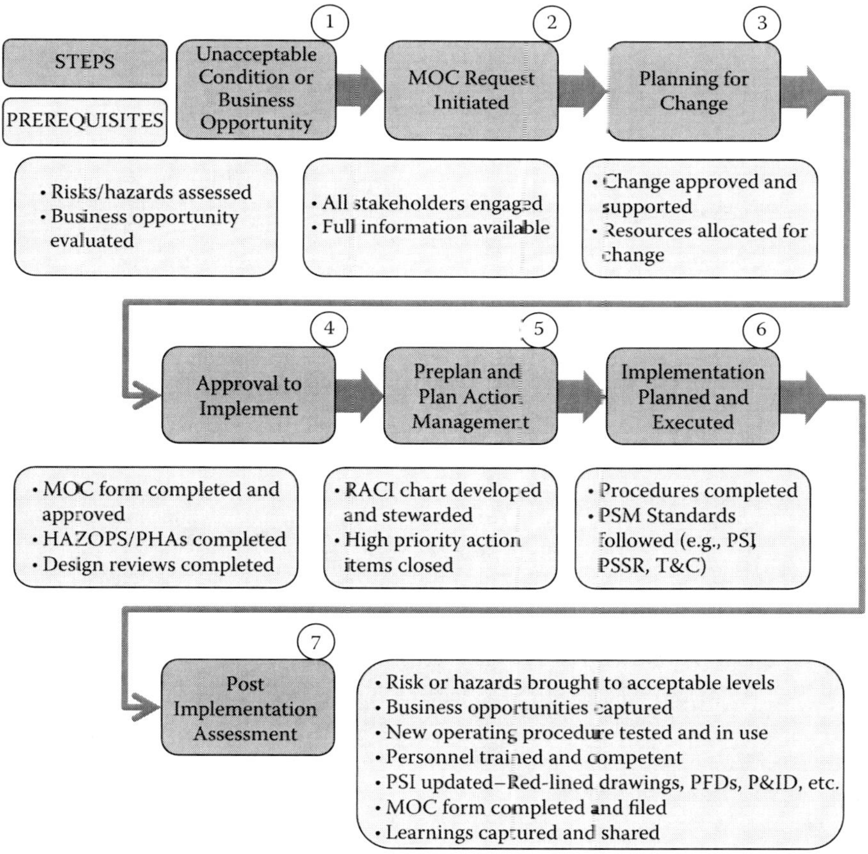

Processes & Systems Element

1. Management of Engineered, Nonengineered, and People Change (MOC)

Management of Change Form

The MOC Notification Form is to contain the following information:

- A description of the change with details about the risk/hazards/opportunity and justification of the proposed change (risk mitigation, financial gain, etc.)
- Initial assessment of current level of residual risk to justify change
- Stakeholder impact assessment – initial assessment of the impact of the change (who will be impacted)
- Modifications to procedures/policies/practices/standards (as applicable)
- Procedure changes, training and communication requirements (as applicable)
- Type of change (engineered, administrative/policy, business procedural, etc.)
- Expected start and completion date
- Regulatory requirements impacted
- Initial assessment of whether process hazard analysis should be considered
- Status category (temporary, emergency, etc.)
- Authorization requirements for the proposed change (based on risk ranking)

Processes & Systems Element

1. Management of Engineered, Nonengineered, and People Change (MOC)

Management of Engineered Change (MOC–E)

- A MOC-E refers to any change:
  - That is a specified addition, alteration, or removal of equipment, facilities, infrastructure, or software
  - To standards or specifications that may result in new components, materials, processes, or procedures being introduced
  - That includes any alterations to the process safety information for a hazardous process

Management of Nonengineered Change (MOC–NE)

- A MOC-NE refers to any change:
  - Of process control systems or operations
  - Of graphic configurations
  - Of equipment titles, descriptions, equipment, and similar related items where engineered specifications or process safety information are not affected and include:
    - Subtle changes that are not replacement-in-kind
    - Changes to assets or software potentially impacting hazardous

Processes & Systems Element

1. Management of Engineered, Nonengineered, and People Change (MOC-P)

- MOC-P outlines requirements for managing *personnel* change to minimize risk introduced by these changes. MOC-P ensures that minimum levels of specific, direct, process experience and minimum levels of knowledge and skill in managing process safety are maintained. MOC-P includes both permanent and temporary changes. It applies to *Business Critical Roles.*

Permanent Changes

- Promotions
- Internal transfers
- Turnover and attrition
- Dismissals

Temporary Changes

- Special assignments
- Approved absenteeism – sick leave and entitlements
- Vacations and extended leaves

Processes & Systems Element

1. **Management of Engineered, Nonengineered, and People Change (MOC)**

Management of Engineered Change

- With Engineered Changes come risks and hazards
  - ➢ Consider each change in a holistic manner
    - ➢ We tend to focus on the small change being introduced as opposed to the change impact to the entire process, facility, or asset
  - ➢ MOC tends to be a common thread across most catastrophic incidents investigated by the Chemical Safety Board

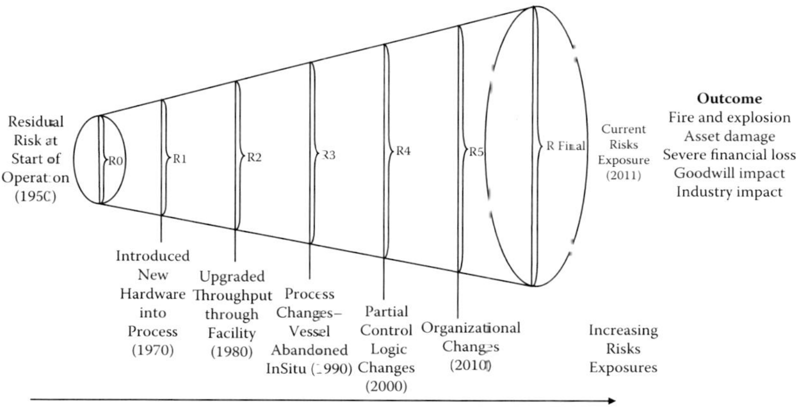

Source: Lutchman, Maharaj, & Ghanem (2012)

Processes & Systems Element

1. **Management of Engineered, Nonengineered, and People Change (MOC)**

Risk Management & the Risk Matrix

- Risk assessment and mitigation is a key MOC requirement
- A risk matrix helps to determine the risk exposure of the organization
- Classifies risks based on probability of occurrence and severity of event or potential event

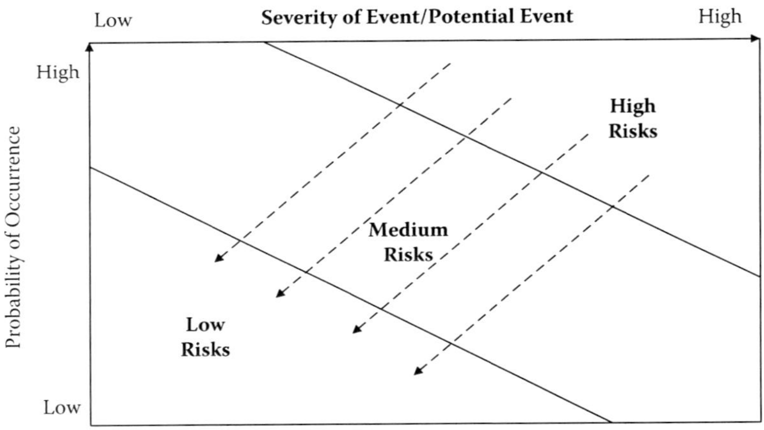

Source: Lutchman, Maharaj, & Ghanem (2012)

Processes & Systems Element

1. **Management of Engineered, Nonengineered, and People Change (MOC)**

Control of Hazards

- Leverages the learnings of the Swiss Cheese Model
- 1st Layer – Design out the risks or hazards
- 2nd Layer – Engineered controls
- 3rd Layer – Administrative controls
- 4th Layer – LAST LINE OF DEFENSE – PPE

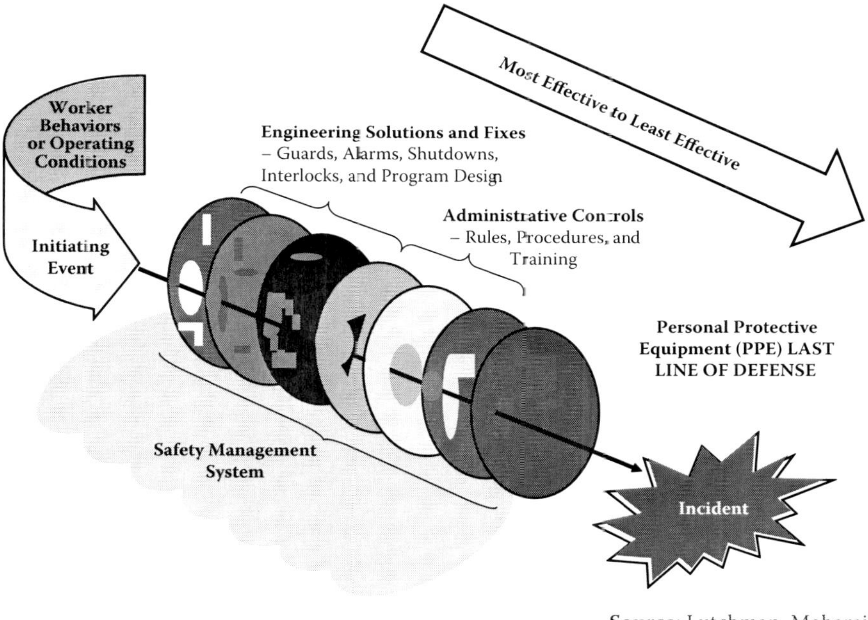

Source: Lutchman, Maharaj, & Ghanem (2012)

Processes & Systems Element

2. Stakeholder Management

Outlines the framework for a systematic approach to the management of communications and stakeholder relations with employees and external stakeholders

Communication

- Addresses the independent needs of both internal and external stakeholder groups
- Credible and honest
- Proactive and transparent

Stakeholder Relations

- Feel and know their concerns are being addressed and interests are being looked after
- Fair treatment in business relationships (contractors and materials suppliers)
- Transparency
- Equal opportunity
- Consultation – unions
- Informed – regulators
- Know what is expected of them
- Collaborative relationships

Processes & Systems Element

2. Stakeholder Management

Stakeholder Interests Varies

- Create a stakeholder interest map
- Understand each stakeholder's interest
- Engage and communicate
- **Not every stakeholder needs to be satisfied ... but they all want to feel considered and engaged**

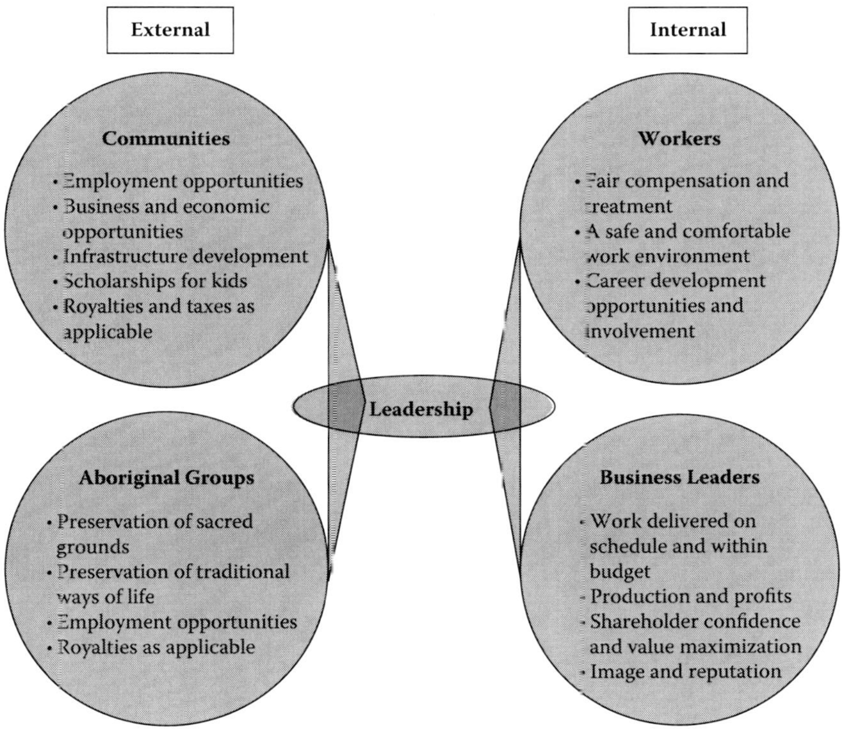

Processes & Systems Element

2. Stakeholder Management

Stakeholder Impact Mapping Critical During Changes

Mapping Requirements

- A clear definition of the desired change
- Stakeholders identified – "Who is affected?"
- Engagement and involvement of stakeholders
- Stakeholder impact assessment – "How are they affected?"
    - Change impact by category
        - Roles and responsibilities
        - Process
        - Technology
        - Skills and knowledge requirements
- Timing of change – "When does change occur?"
    - Plan for periods of minimal impact to stakeholders
- Support required in executing change
    - Communication
    - Training and competency assurance
    - Leadership alignment
- Risk associated with executing the change
    - Risk mitigation and action management plans

Processes & Systems Element

3. Risk Management

Risk Management Expectations

Establishes expectations for the implementation of a systematic approach to identify and manage risk through the use of standardized tools and processes

- All work should be risk ranked before work is undertaken
- No work will be done in situations of unacceptable risk levels
  - Risks must be mitigated to acceptable levels before work can be undertaken
- A risk matrix should be accessible to all operations and maintenance work groups

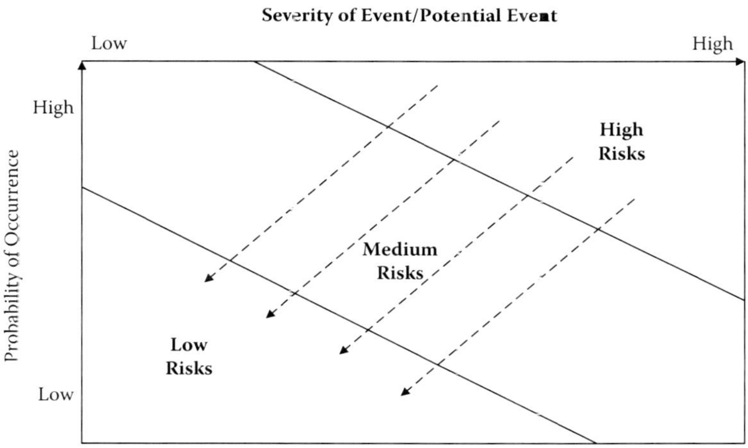

Source: Lutchman, Maharaj, & Ghanem (2012)

Processes & Systems Element

4. Prestart-up Safety Reviews (PSSR)

Establishes expectations for the implementation of a systematic approach to review the readiness of a facility or operation before start-up

- Incidents generally occur after a change is made to operating conditions
  - Engineering changes (addition or removal of hardware)
  - Facilities or plant turnaround

PSSR Team

Prior to the introduction of hazardous materials or energy to new or modified assets, a multidisciplinary prestart-up safety review team shall be constituted to verify that

- Process hazards recommendations required before start-up are complete
- Tests and inspections are complete
- General requirements for safety, health, and environmental considerations are adequate – such as fire protection, ergonomics, environment, and physical conditions
- Operating and maintenance procedures have been written and authorized
- Training requirements are complete
- Construction and equipment are in accordance with design specifications
- The process safety management elements were addressed to ensure changes have been properly reflected in those elements
- A physical inspection was conducted in the field guided by pre-written checklists

Processes & Systems Element

4. Prestart-up Safety Reviews (PSSR)

PSSR Provides For

- A cold-eye review of facilities via the following:
  - ➢ Systems walk-down to ensure system integrity
  - ➢ Flanges bolted properly
  - ➢ Drains appropriately closed
  - ➢ Gauges installed as required
  - ➢ Blinds and isolations removed
- Process Safety Information (PSI) verification
  - ➢ Standard operating procedures
  - ➢ MOC Compliance
  - ➢ P&ID, PFDs
- Training and competency verification
  - ➢ Training matrix completed
  - ➢ Critical personnel demonstrate competency in ability to execute work
- Regulatory compliance
  - ➢ Verification that facility meets or exceeds regulatory requirements
- Stakeholder engagement and communication
  - ➢ Stakeholder impact assessment mapping review
- Action management review and verification
  - ➢ Confirmation that all high-risk action items are closed

Processes & Systems Element

5. Audits & Assessments

Organizations must establish clear requirements for the implementation and maintenance of audit and assessment processes at various levels of the organization. Goal of the Audit and Assessment Organization is to assist the organization in proactively identifying gaps the organization's PSM, SMS, and OMS systems to prevent incidents and to maximize value of the organization.

Success of audits and assessments lies in *avoiding the blame game. Focus on collaboration and proactive solutions.*

Early Deming Work on Root Causes of Incidents

- 85% – Related to inappropriate and ineffective processes
- 15% – Originates from people

Source: Hein (2003)

Types of Audits

- Process Safety Management (PSM)
- Safety Management System (SMS)
- Operations Management System (OMS)
- Internal Self-Assessment
- Corporate Audit Team
- External Auditors
- Contractor Audits

Processes & Systems Element

5. Audits & Assessments

Audit Teams Expertise and Composition

- Audit Team Leader
- Subject Matter Experts (SMEs)
- Operations Management
- EH&S/HSSE/HSEQ

Audit Preparation

- Pre-audit Preparation (1–2 months before the audit)
    - ➢ Ensure leadership buy-in and support
    - ➢ Work with business to schedule and support audit process
    - ➢ Ensure business has access to audit protocol
    - ➢ Review prior audits if completed and high-priority action items from prior audit
    - ➢ Business to highlight any areas of special interest and audit focus
    - ➢ Auditors must be prepared and ready for the audit
    - ➢ Business stakeholders participating in the audit must be engaged and scheduled
    - ➢ Avoid disrupting the business and its operations
    - ➢ Put stakeholders at ease – avoid witch hunts!!!

Audit Tool (if available)

- Preload audit protocol requirement for each Standard or Element being audited
- Ensure all users have access to the tool and are trained in use of use of the tool
- Where applicable, BU stakeholder (SME) to begin documentation mapping to demonstrate compliance to standard/element

Processes & Systems Element

5. Audits & Assessments

Audit Focus

All safety audit comments, recommendations, and corrective actions should focus on the business responses to these four questions:

- o Does the business processes meet all regulatory and industry best practices requirements?
- o Are the SMS requirements being met?
- o Is there documented proof of compliance?
- o Is employee training effective – can and do they apply specific safe behaviors?

Steps in Conducting the Audit

1. Opening meeting – auditors and business stakeholders
    - ➢ Clearly identify goals and objectives of the audit
    - ➢ Reconfirm audit is collaborative and intended to proactively assist the business in identifying and closing high-risk gaps
2. Audit or fact-finding exercise
    - ➢ Interviews of stakeholders
    - ➢ Documentation reviews
    - ➢ System walk-downs
        - ➢ Equipment maintenance
        - ➢ Housekeeping
        - ➢ People
    - ➢ Engaging the workforce: are they aware of procedures, emergency response, muster points, work practices, etc.?
    - ➢ *Trust ... but verify*

Processes & Systems Element

5. Audits & Assessments

Steps in Conducting the Audit (continued)

3. Compare findings with procedure, standard, and element requirements
   - ➢ Identify gaps in compliance and areas of best practices
   - ➢ Identify low-hanging fruits and high-risk exposures
   - ➢ Risk rank and prioritize gap closure opportunities
   - ➢ Identify corrective actions and recommendations for gap closures
   - ➢ Work with business stakeholders to score the business compliance relative to compliance requirements
     - ➢ Ensure business ownership of the score – gain consensus on the scoring

4. Debrief with stakeholders and business leadership
   - ➢ *Start with the good things being done*
   - ➢ Identify the low-hanging fruits that require little effort and resources to close the gaps
   - ➢ Review prioritized high-risk opportunities
   - ➢ Review corrective actions and recommendations for gap closure strategies
   - ➢ Wrap up by recapping and reaffirming support on corrective actions and recommendations

5. Follow-up and business unit support
   - ➢ Action management stewardship and follow-up

Processes & Systems Element

6. Management Reviews

Establish and Steward SMART Performance Indicators

Leadership establishes the requirements reporting on Key Performance Indicators (KPIs) to ensure continuous improvements and to assess the efficacy of processes and tools that may be in place to support the OMS

1. Establishing KPIs
   - ➢ Leading and lagging indicator focus
   - ➢ Emphasis on leading indicators
   - ➢ Monthly reporting and stewardship

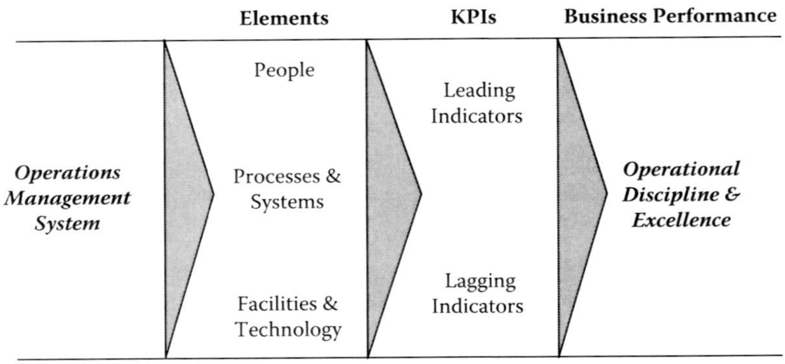

Processes & Systems Element

6. Management Reviews

Operations Management System

People Elements

Leadership, management, and organizational commitment

Security management and emergency preparedness

Qualification, orientation, and training

Contractor/supplier management

Event management and learning

Goals, targets, and planning

Facilities & Technology Elements

Documentation management and process safety information

Process hazard analysis (PHA)

Physical asset system integrity and reliability

Operating procedure and safe work practices

Legal requirements and commitment

Processes & Systems Elements

Management of engineered, non-engineered, and people change

Stakeholder management and communications

Risk management

Prestart-up safety reviews (PSSR)

Audits & assessments

Management review

Approaches to KPIs

- o Identify 3–5 KPIs for each element for stewardship
- o Select high-value KPIs to focus on
- o Focus on a maximum of 2–3 KPIs to steward
- o Ensure SMART capabilities for each KPI
- o Determine the right frequency for reporting on KPIs – monthly vs. quarterly
- o Avoid taxing the business in reporting requirements

SMART KPIs

People Elements	Leading Indicators	Lagging Indicators
Leadership, management, and organizational commitment	o Number of field visits completed o Number of communication sessions completed	o N/A
Security management and emergency preparedness	o Number of self-assessment audits completed o Number of security gaps identified o Number of emergency management drills completed vs. scheduled	o Number of security breaches o Number of gaps closed o Number of severe incidents requiring emergency response activation
Qualification, orientation, and training	o % of workforce competency assured o % of workforce trained o % of contractors oriented before arriving to site	o Number of incidents resulting from deficiencies in training and competency
Contractor/supplier management	o Number of joint contractor /owner leadership visits to the frontline o Number of contractor audits completed	o CRIF, CDIF o Number of recordable injuries o Number of contractor fatalities
Event management and learning	o Number of high-priority incidents investigated o % of corrective actions implemented and closed o Average age of corrective actions	o Number of high-priority incidents o Number of incidents
Goals, targets, and planning	o % of work group with completed work plans o Number of performance evaluations completed	o % of work group without work plans

SMART KPIs

Facilities &Technology Elements	Leading Indicators	Lagging Indicators
Documentation management and process safety information		
Process hazard analysis (PHA)	o Number of PHAs past cycle requirements	
Physical asset system integrity and reliability	o Overdue tests and inspections of PS-critical equipment o Number of bypassed interlocks o Number of bypassed alarms	
Operating procedure and safe work practices	o Number of spot audits done on procedure use and safe work practices	o Number of incidents related to absence of procedure and safe work practices
Legal requirements and commitment	o Number of regulatory incompliance	o Number of litigation cases

SMART KPIs

Processes & Systems Elements	Leading Indicators	Lagging Indicators
Management of engineered, nonengineered, and people change	o Total number of temporary changes o Total number of temporary changes past due dates	o N/A
Stakeholder management and communications	o Number of stakeholder engagement sessions conducted	o Number of protests
Risk management	o % of work permits reviewed with incorrect hazards assessments	o Number of incidents arising from incorrect hazards assessments and risk ranking
Prestart-up safety reviews (PSSR)	o Number of start-ups completed without a documented PSSR	o N/A
Audits & assessments	o Number of self-assessments completed o % of audits not completed vs. scheduled	o Number of incidents resulting from identified audit gaps
Management review	o Number of corrective actions issued related to KPIs gaps	o Number of KPIs not being met o Number of KPIs not being followed up on

Fundamental 3

Establishing the Baseline

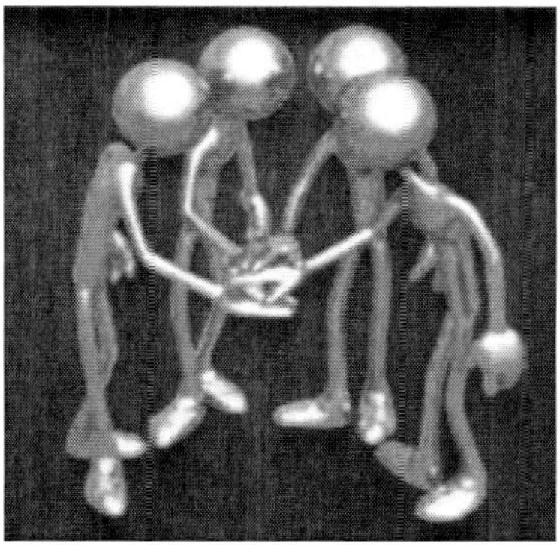

In this section you will gain insights into

1. Planning for establishing the baseline
2. Gathering relevant information to confirm compliance
3. Independent verification of supporting information
4. Gaps identification and implementation plan

Discussion Topics

- Planning for Establishing the Baseline
- Documentation Mapping and Information Gathering
- Verification of Management System Maturity
- Implementation Plan Development
- Implementation and Sustainment – PDCA Model

Elements: People, Processes & Systems, and Facilities & Technology

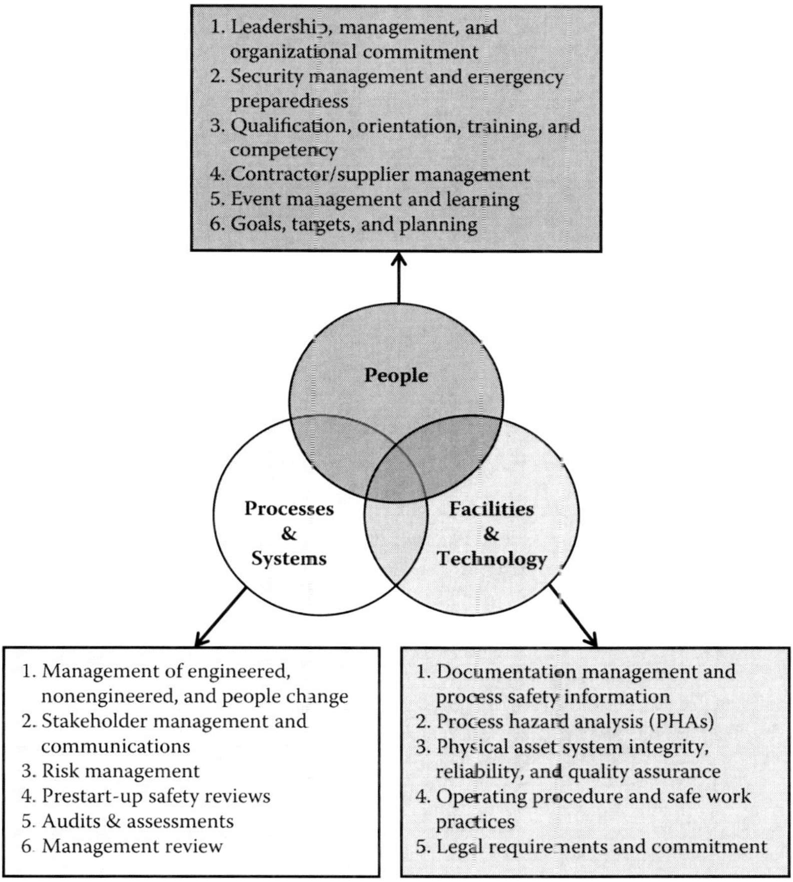

Understanding the Element Compliance Process

Element
Event Management and Learning

Subelement Requirements

- o Incidents and hazards reporting
- o Incidents and hazards risk assessments
- o Incident investigations and root cause analysis
- o Corrective actions management
- o Knowledge generated and Learning from Events (LFE) process

Implementation Guidelines

- o Consistent interpretation of each subelement requirement

Subelement Interpretation

- o Incidents and hazards reporting – A process/tool is in place for all personnel to report and record hazards and incidents. Personnel are trained in use of the process/tool. The process/tool provides a notification for all high-risk incidents and hazards with appropriate tracking mechanisms to ensure action management and follow-up

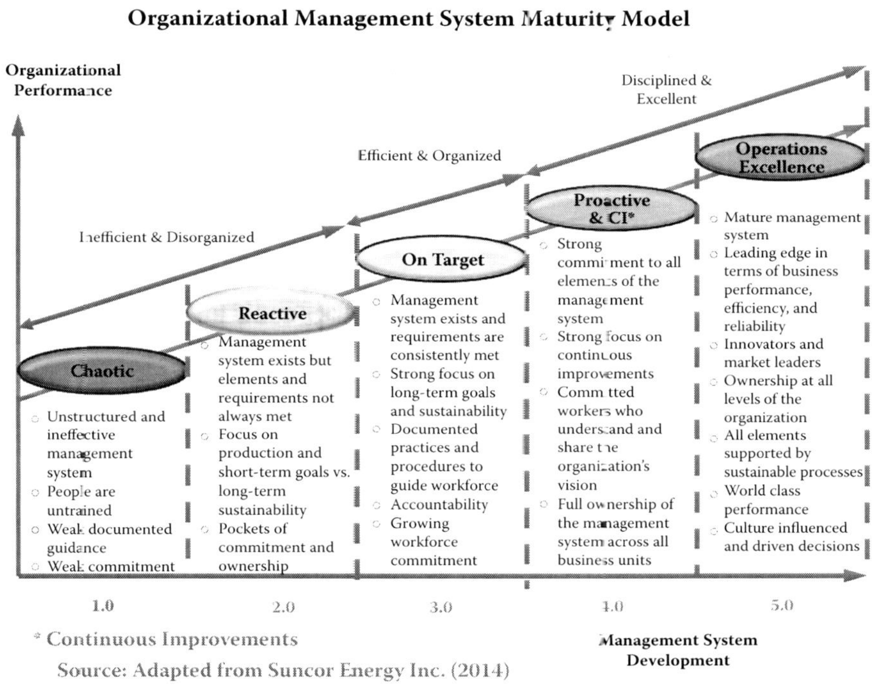

Source: Adapted from Suncor Energy Inc. (2014)

- Each element requirement must be accompanied by an implementation guideline to ensure consistent interpretation and field application
- Applicable scores are defined by the level of compliance to the element requirements
- At the very minimum, the organization must strive for ***Planned*** *Compliance and Development Levels*

Steps in Establishing the Baseline

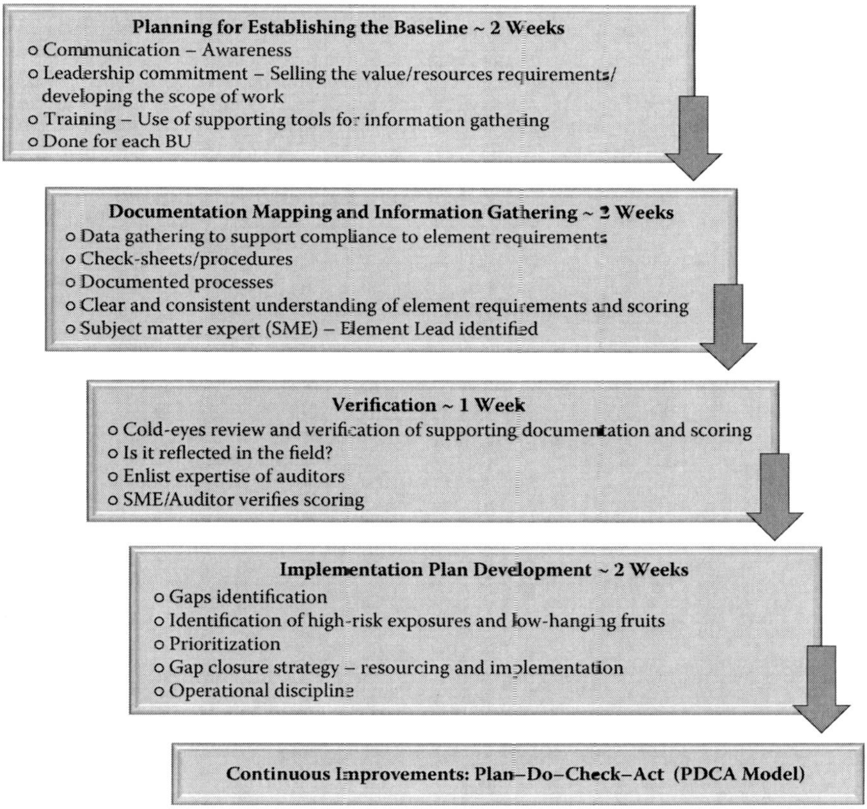

Planning for Establishing the Baseline

Planning for Establishing the Baseline ~ 2 Weeks

- Communication – Awareness
- Leadership commitment – Selling the value/resources requirements/ developing the scope of work
- Training – Use of supporting tools for information gathering
- Done for each BU

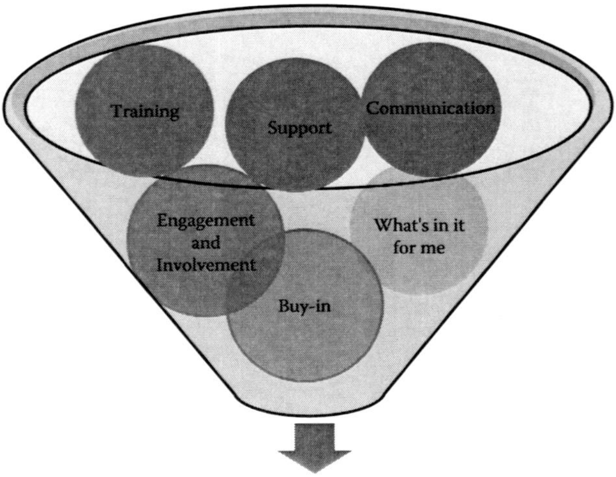

Leadership Commitment for Baseline Assessment

Documentation Mapping and Information Gathering

Documentation Mapping and Information Gathering ~ 2 Weeks

- Data gathering to support compliance to element requirements
- Check-sheets/procedures
- Documented processes
- Clear and consistent understanding of element requirements and scoring
- Subject matter expert (SME) – Element Lead identified

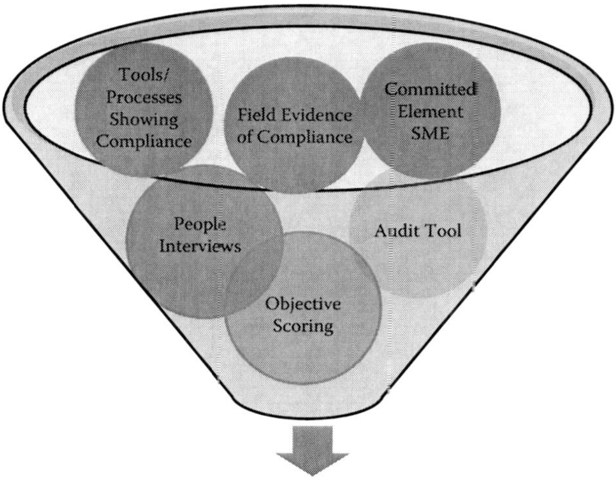

Self-Assessed Management System Maturity Score

Verification

Verification ~ 1 Week

- Cold-eyes review and verification of supporting documentation and scoring
- Is it reflected in the field?
- Enlist expertise of auditors
- SME/Auditor verifies scoring

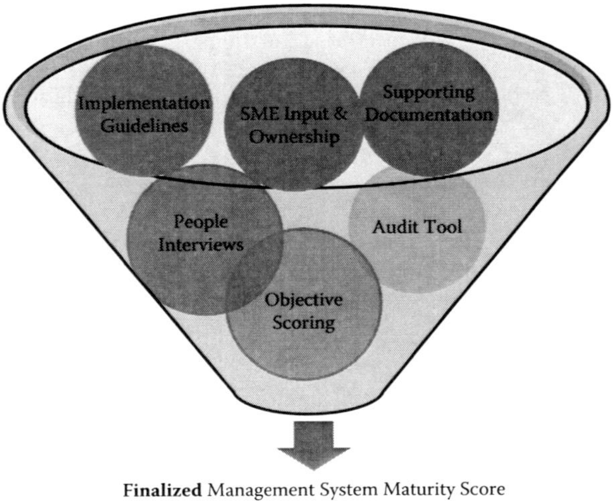

Implementation Plan Development

Implementation Plan Development ~ 2 Weeks

- Gaps identification
- Identification of high-risk exposures and low hanging fruits
- Prioritization
- Gap closure strategy – resourcing and implementation
- Operational discipline

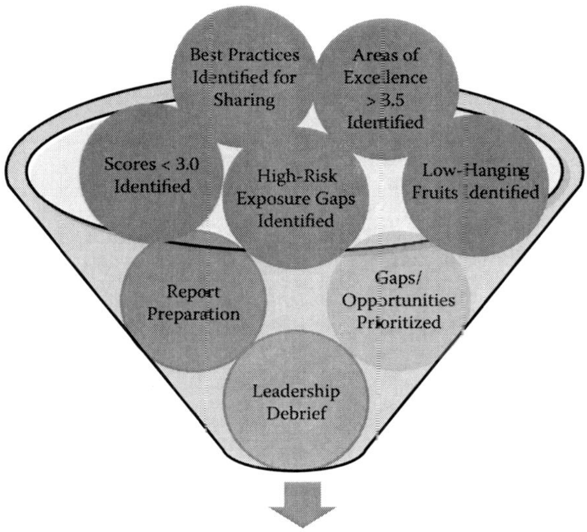

Implementation Plan Development

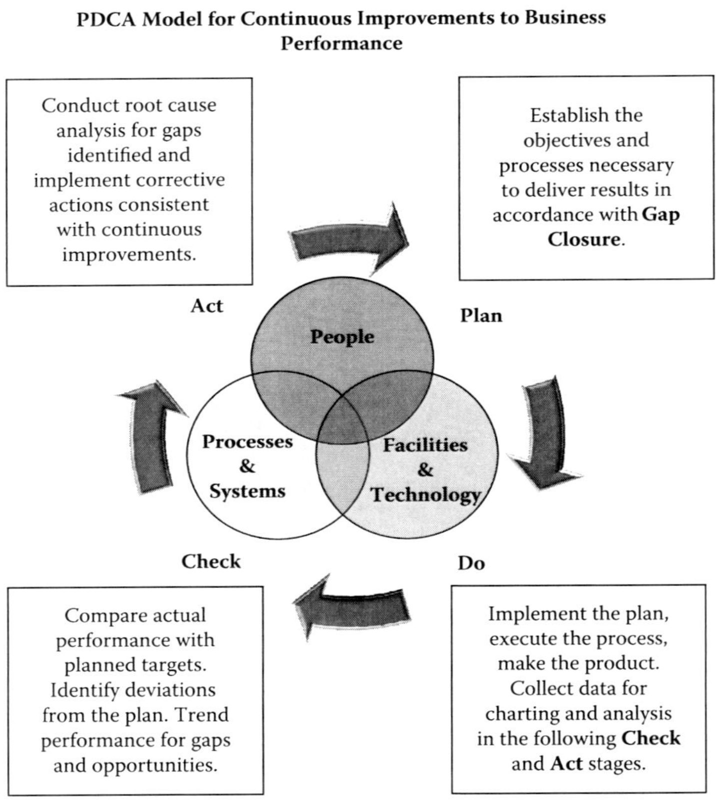
PDCA Model for Continuous Improvements to Business Performance
Conduct root cause analysis for gaps identified and implement corrective actions consistent with continuous improvements.
Establish the objectives and processes necessary to deliver results in accordance with Gap Closure.
Act
Plan
People
Processes & Systems
Facilities & Technology
Check
Do
Compare actual performance with planned targets. Identify deviations from the plan. Trend performance for gaps and opportunities.
Implement the plan, execute the process, make the product. Collect data for charting and analysis in the following Check and Act stages.

Fundamental 4

Plan–Do–Check–Act

In this section you will gain insights into

1. Planning
2. Doing – Implementation
3. Checking – Evaluation and analysis
4. Acting – Following up

Discussion Topics

- The PDCA Model
- Planning
- Doing
- Checking
- Acting

Whether your objectives have been given to you by your boss or you're creating objectives, make sure they are SMART—specific, measurable, achievable, realistic, and time bounded.

Elements: People, Processes & Systems, and Facilities & Technology

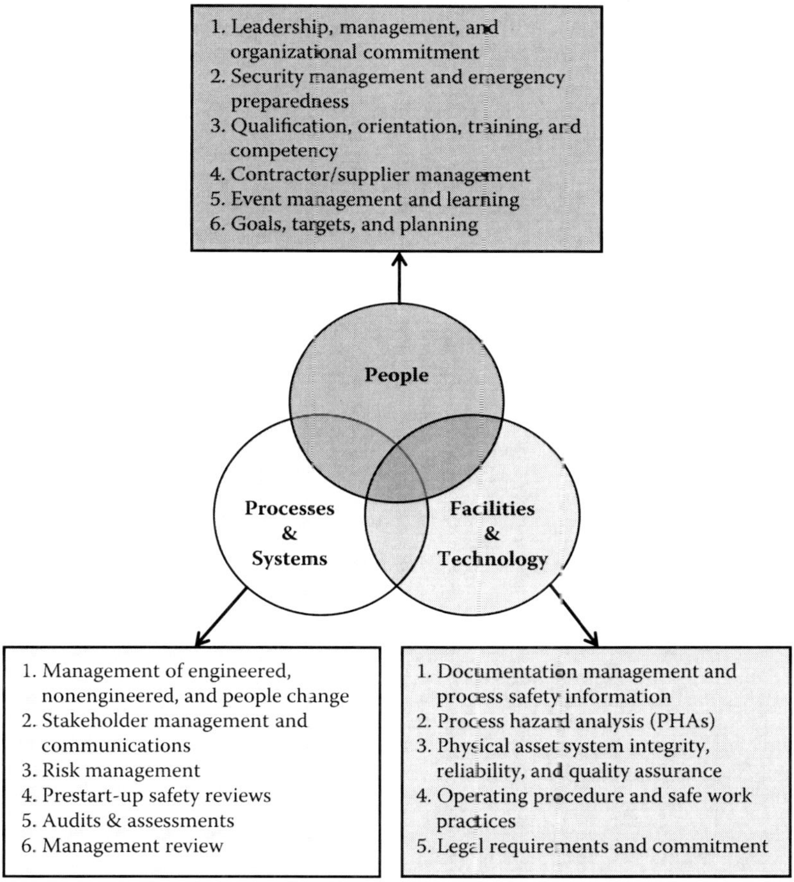

PDCA Model for Continuous Improvements to Business Performance – Deming (1950s)

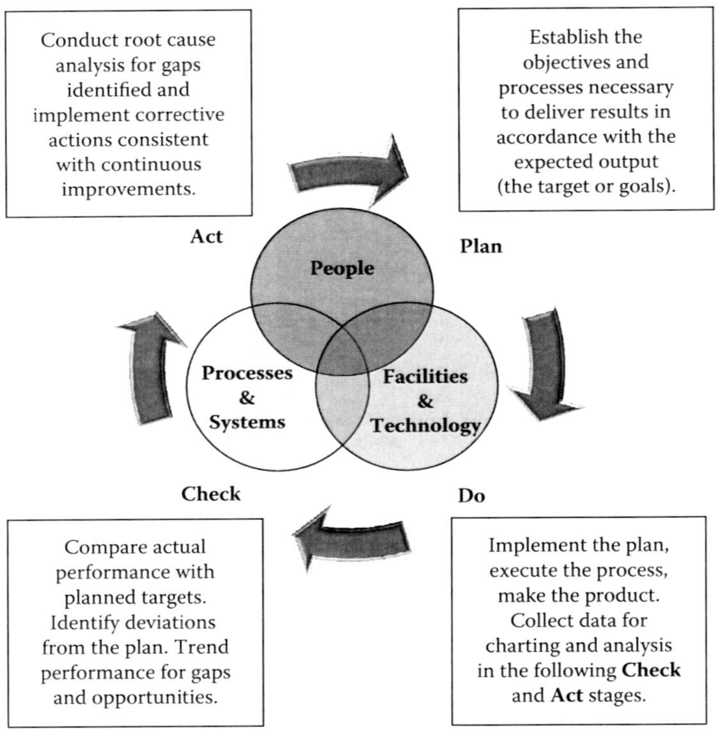

Planning

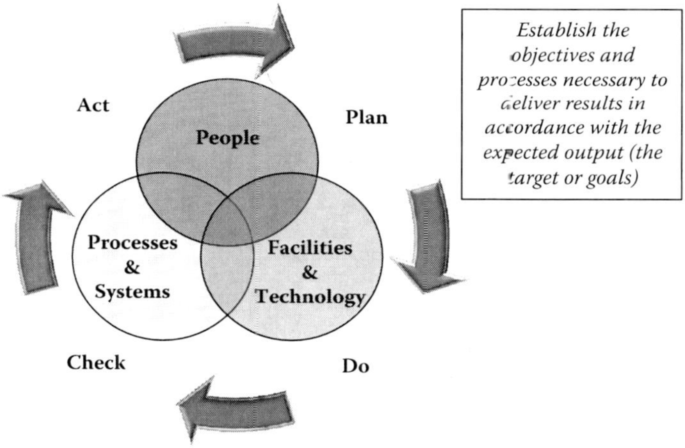

- Creating objectives, goals, and targets that are SMART—specific, measurable, achievable, realistic, and time bound
- Develop strategies for achieving objectives, goals, and targets
- Scheduling and milestones identification

Doing – Implementation

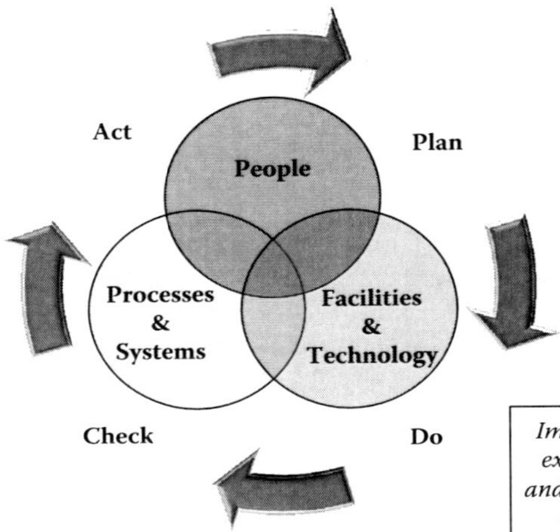

*Implement the plan, execute the process, and make the product. Collect data for charting and analysis in the following **Check** and **Act** stages:*

- Resources allocation – People, $$$$, leadership, materials
- Follow the schedule to achieve milestones
- Readiness reviews
- Continuous operations
- Data collection and trending
- Compliance to Standards, Procedures, and OEMS Elements
- Compliance to legal and regulatory requirements

Checking – Evaluation and Analysis

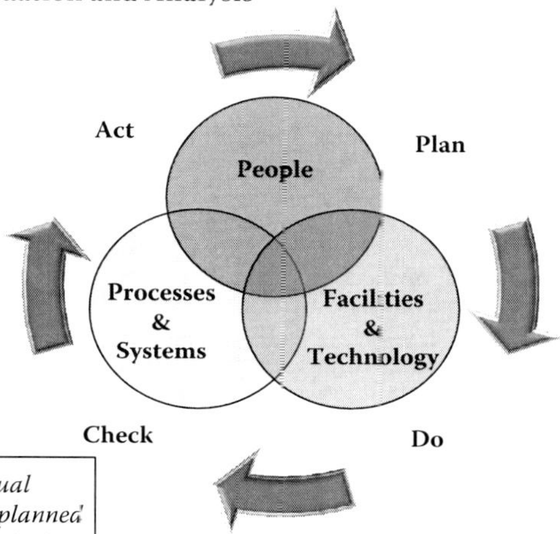

Compare actual performance with planned targets. Identify deviations from the plan. Trend performance for gaps and opportunities.

- Key Performance Indicators (KPIs) stewardship
- Graphs and trending
- Self-assessments and audits
  - Performance comparison against Standards and Procedures
  - Internal and external audits
  - Root cause analysis (RCA)
- Identify and prioritize gaps based on risks an value creation/maximization opportunities

Acting – Following up

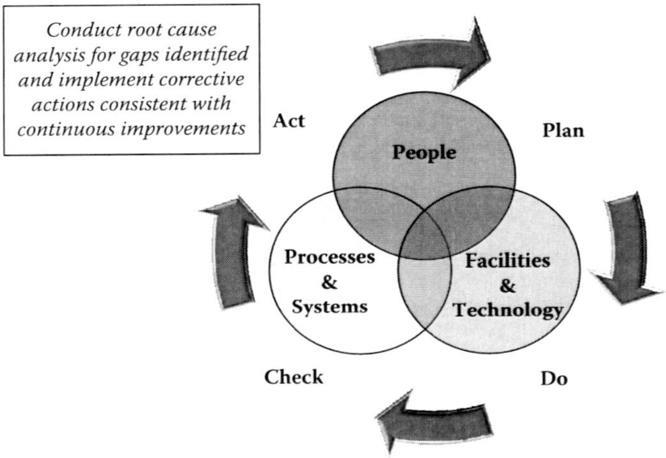

- Identify and prioritize gaps based on risks and value creation / maximization opportunities
  - 1st priority – High-risk gaps
  - 2nd priority – Low-hanging fruits
- Implement corrective actions and gap closure strategies
- Ensure *Leadership commitment*

Establish and Steward SMART Performance Indicators

Leadership establishes the requirements reporting on Key Performance Indicators (KPIs) to ensure continuous improvements and to assess the efficacy of processes and tools that may be in place to support the OMS

1. Establishing KPIs
   - ➢ Leading and lagging indicator focus
   - ➢ Emphasis on leading indicators
   - ➢ Monthly reporting and stewardship

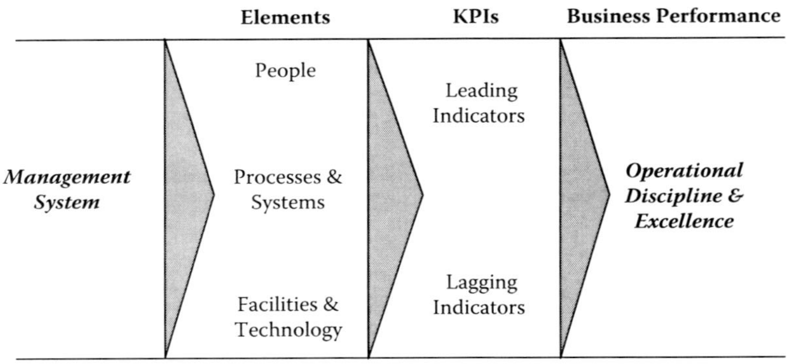

Operations Management System

People Elements

Leadership, management, and organizational commitment

Security management and emergency preparedness

Qualification, orientation, and training

Contractor/supplier management

Event management and learning

Goals, targets, and planning

Facilities & Technology Elements

Documentation management and Process safety information

Process hazard analysis (PHA)

Physical asset system integrity and reliability

Operating procedure and safe work practices

Legal requirements and commitment

Processes & Systems Elements

Management of engineered, non-engineered, and people change

Stakeholder management and communications

Risk management

Prestart-up safety reviews (PSSR)

Audits & assessments

Management review

Approaches to KPIs

Leading **Lagging**

- Identify 3–5 KPIs for each element for stewardship
- Select high-value KPIs to focus on
- Focus on a maximum of 2–3 KPIs to steward
- Ensure SMART capabilities for each KPI
- Determine the right frequency for reporting on KPIs – monthly vs. quarterly
- Avoid taxing the business in reporting requirements

SMART KPIs

People Elements	Leading Indicators	Lagging Indicators
Leadership, management, and organizational commitment	o Number of field visits completed o Number of communication sessions completed	o N/A
Security management and emergency preparedness	o Number of self-assessment audits completed o Number of security gaps identified o Number of emergency management drills completed vs. scheduled	o Number of security breaches o Number of gaps closed o Number of severe incidents requiring emergency response activation
Qualification, orientation, and training	o % of workforce competency assured o % of workforce trained o % of contractors oriented before arriving to site	o Number of incidents resulting from deficiencies in training and competency
Contractor/supplier management	o Number of joint contractor /owner leadership visits to the frontline o Number of contractor audits completed	o CRIF, CDIF o Number of recordable injuries o Number of contractor fatalities
Event management and learning	o Number of high-priority incidents investigated o % of corrective actions implemented and closed o Average age of corrective actions	o Number of high-priority incidents o Number of incidents
Goals, targets, and planning	o % of work group with completed work plans o Number of performance evaluations completed	o % of work group without work plans

SMART KPIs

Facilities &Technology Elements	Leading Indicators	Lagging Indicators
Documentation management and process safety information		
Process hazard analysis (PHA)	o Number of PHAs past cycle requirements	
Physical asset system integrity and reliability	o Overdue tests and inspections of PS critical equipment o Number of bypassed interlocks o Number of bypassed alarms	
Operating procedure and safe work practices	o Number of spot audits done on procedure use and safe work practices	o Number of incidents related to absence of procedure and safe work practices
Legal requirements and commitment	o Number of regulatory incompliance	o Number of litigation cases

SMART KPIs

Processes & Systems Elements	Leading Indicators	Lagging Indicators
Management of engineered, nonengineered, and people change	o Total number of temporary changes o Total number of temporary changes past due dates	o N/A
Stakeholder management and communications	o Number of stakeholder engagement sessions conducted	o Number of protests
Risk management	o % of work permits reviewed with incorrect hazards assessments	o Number of incidents arising from incorrect hazards assessments and risk ranking
Prestart-up safety reviews (PSSR)	o Number of start-ups completed without a documented PSSR	o N/A
Audits & assessments	o Number of self-assessments completed o % of audits not completed vs. scheduled	o Number of incidents resulting from identified audit gaps
Management review	o Number of corrective actions issued related to KPIs gaps	o Number of KPIs not being met o Number of KPIs not being followed up on

Fundamental 5

Auditing for Compliance and Conformance

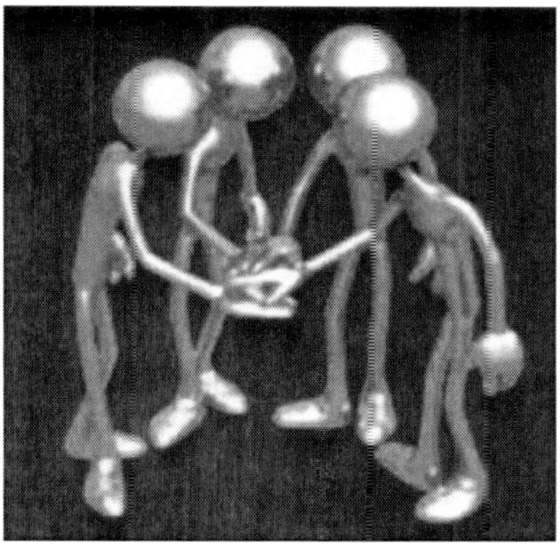

In this section you will gain insights into

1. The audit process
2. Auditing cycle
3. Self-assessments, 2nd & 3rd party auditing

Discussion Topics

- Collaboration vs. confrontation
- The audit process
- Requirements for a successful audit
- Self-assessments and 2nd & 3rd party auditing

Elements: People, Processes, & Systems, and Facilities & Technology

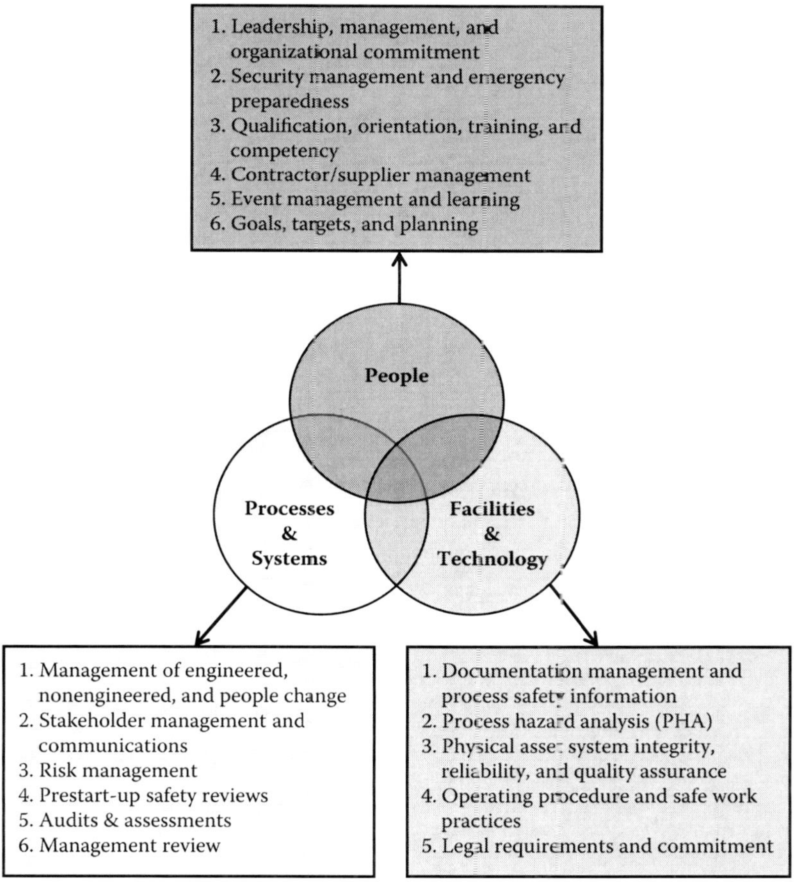

Understanding the Element Compliance Process

Element
Event Management and Learning

Audit Protocol
Identifies what's in scope for the audit

Subelement Requirements

- o Incidents and hazards reporting
- o Incidents and hazards risk assessments
- o Incident investigations and root cause analysis
- o Corrective actions management
- o Knowledge generated and Learning from Events (LFE) process

Implementation Guidelines

- o Consistent interpretation of each subelement requirement

Subelement Interpretation

- o Incidents and hazards reporting – A process/tool is in place for all personnel to report and record hazards and incidents. Personnel are trained in use of the process/tool. The process/tool provides allows for risk ranking and generating notifications for all high-risk incidents and hazards with appropriate tracking mechanisms to ensure action management and follow-up

Organizational Management System Maturity Model

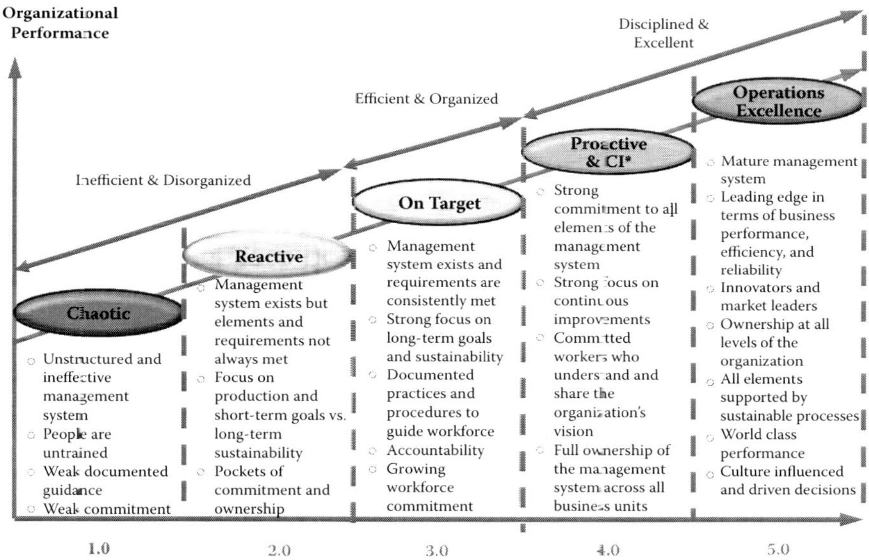

* Continuous Improvements

Source: Adapted from Suncor Energy Inc. (2014)

- Each element requirement must be accompanied by an implementation guideline to ensure consistent interpretation and field application
- Applicable scores are defined by the level of compliance to the element requirements
- At the very minimum, the organization must strive for ***Planned*** *Compliance and Development Levels*

Auditing from the Baseline or Prior Audit

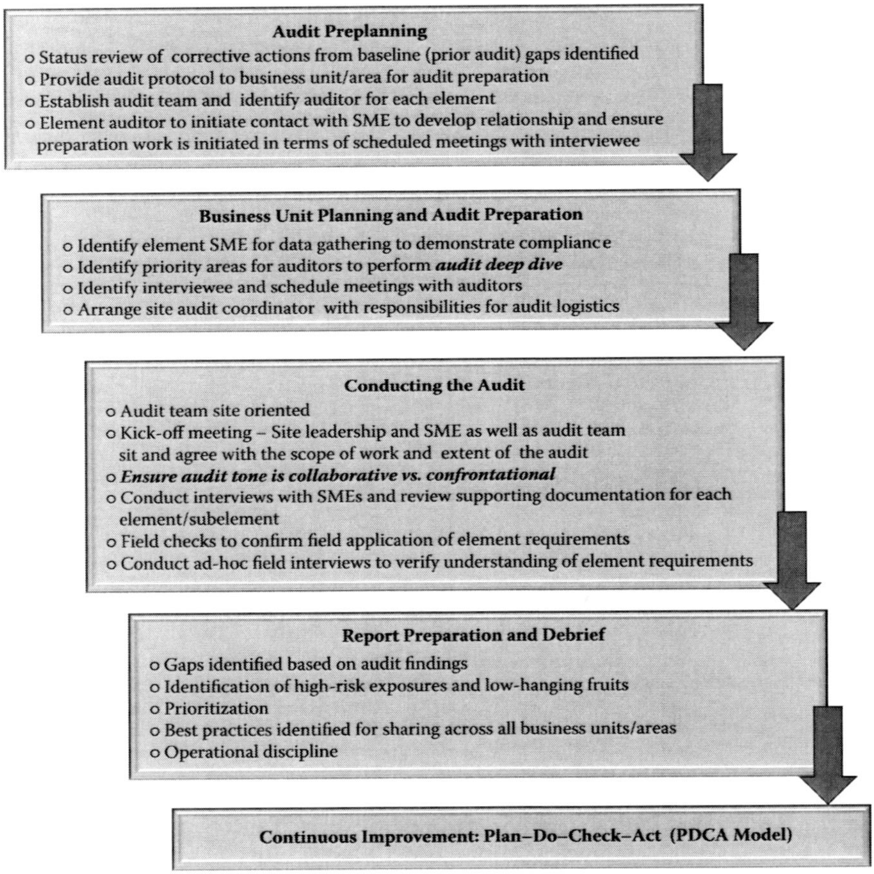

Requirements for a Successful Audit

General Requirements

- Position audit to be a proactive approach to identifying opportunities as opposed to poking holes
- Collaboration – be sure all interviewees and SMEs are fully at ease
- Avoid the blame game
- Do not be a bearer of bad news only – highlight the positive things being done and be liberal in praises for good work

Audit should answer the following questions:

- Does the audit cover all regulatory and best industry practice requirements?
- Are the requirements of each element being met?
- Is there documented proof of compliance?
- Is employee training effective – can and do they apply specific safe behaviors?

Auditors Requirements

- Auditors must possess the right level of audit experience
- Comprehensive experience in compliance, including regulatory requirements
- A strong knowledge base including relevant industry knowledge where applicable
- Proper accreditation for the audit operations required

Types of Audit

Auditing

- **Self-Assessments** – Auditors and audit team are internal to the business unit
- **2nd Party Audits** – Auditors and audit team are sourced from an alternative business unit. For example, BU 1 audit can be done by auditors from other, e.g., BU 2-3
- **3rd Party Audits** – Auditors external to the organization are contracted to perform the business unit audit

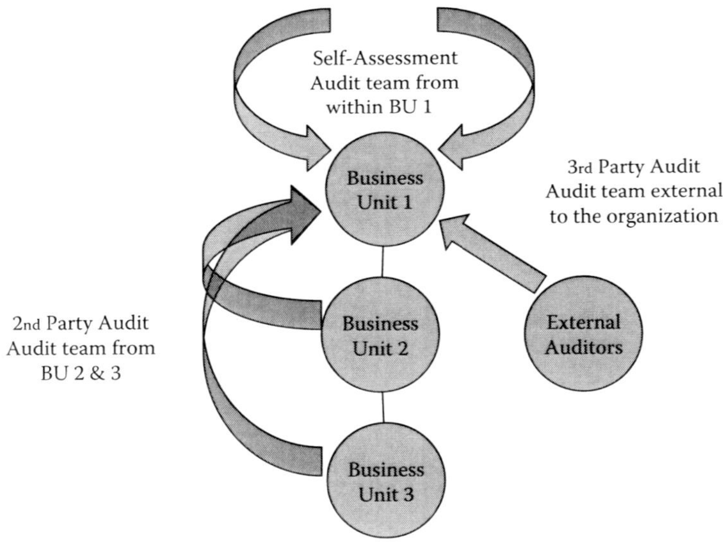

Audit Findings

Your credibility depends on the findings you determine as an auditor

Relevant and Applicable

- Findings must be applicable to the process being audited

Measurable

- Can we quantify the gap?

Material

- 3 out of 270 procedures were not updated
- 40 out of 200 work permits were not properly filled out

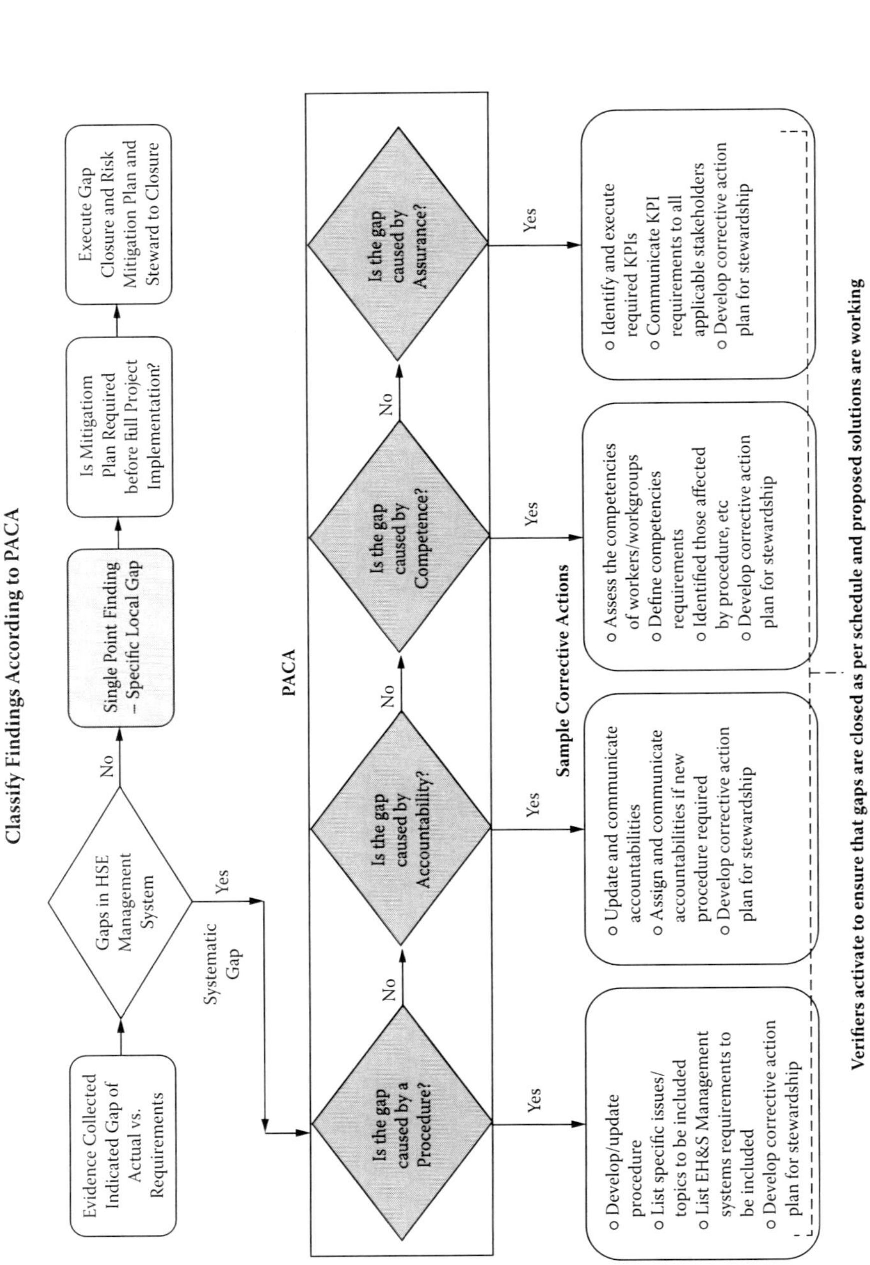

Classify Findings According to PACA
Evidence Collected Indicated Gap of Actual vs. Requirements
Gaps in HSE Management System
No
Single Point Finding – Specific Local Gap
Is Mitigatiom Plan Required before full Project Implementation?
Execute Gap Closure and Risk Mitigation Plan and Steward to Closure
Yes
Systematic Gap
PACA
Is the gap caused by a Procedure?
No
Is the gap caused by Accountability?
No
Is the gap caused by Competence?
No
Is the gap caused by Assurance?
Yes
Yes
Yes
Yes
Sample Corrective Actions
o Develop/update procedure
o List specific issues/ topics to be included
o List EH&S Management systems requirements to be included
o Develop corrective action plan for stewardship
o Update and communicate accountabilities
o Assign and communicate accountabilities if new procedure required
o Develop corrective action plan for stewardship
o Assess the competencies of workers/workgroups
o Define competencies requirements
o Identified those affected by procedure, etc
o Develop corrective action plan for stewardship
o Identify and execute required KPIs
o Communicate KPI requirements to all applicable stakeholders
o Develop corrective action plan for stewardship
Verifiers activate to ensure that gaps are closed as per schedule and proposed solutions are working

Audit Scores and Scoring

Auditing

- Auditors (SMEs) interpret requirements for compliance to the element requirements and allocate a score for each subelement of the management system
- Interpretation guidelines used to drive consistency in interpretation across all business units/business areas

Scoring

Method 1: Weighted Average

- Element Score 0–5
  - Subelement Score 0–5

Σ Weighted Average of Subelement Scores = Element Score

Σ Weighted Average of Element Scores = BA Audit Score

Method 2: Average

- Element Score 0–5
  - Subelement Score 0–5

Σ Average of Subelement Scores = Element Score

Σ Average of Element Scores = BA Audit Score

Audit Scores and Scoring

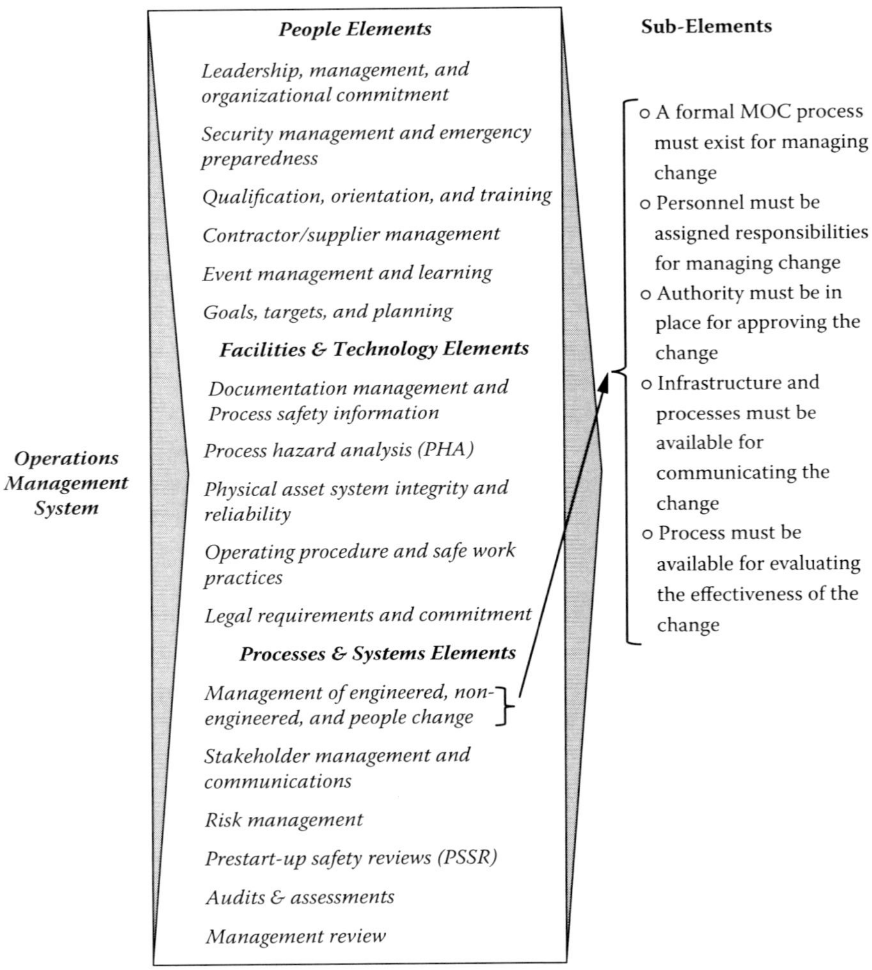

PDCA Model for Continuous Improvements to Business Performance

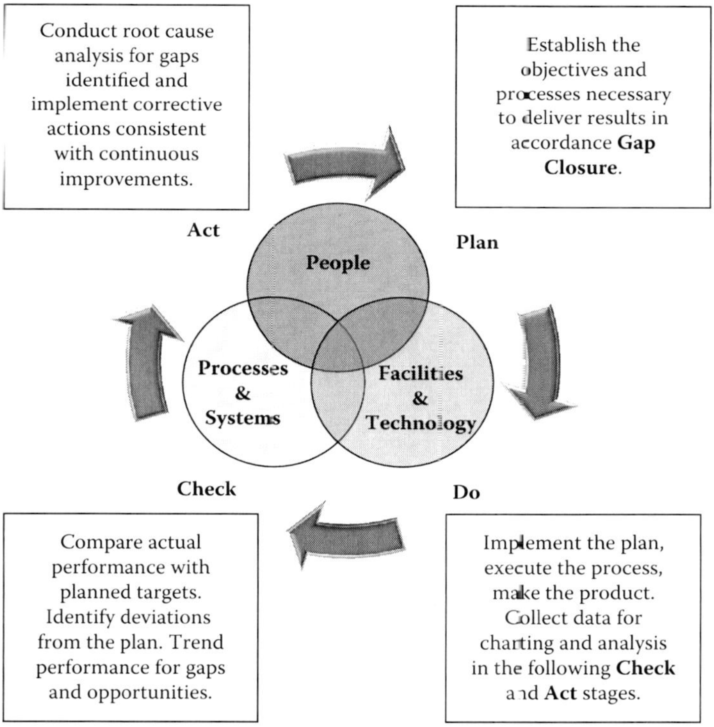

Fundamental 6

Closing the Gaps – Operations Discipline

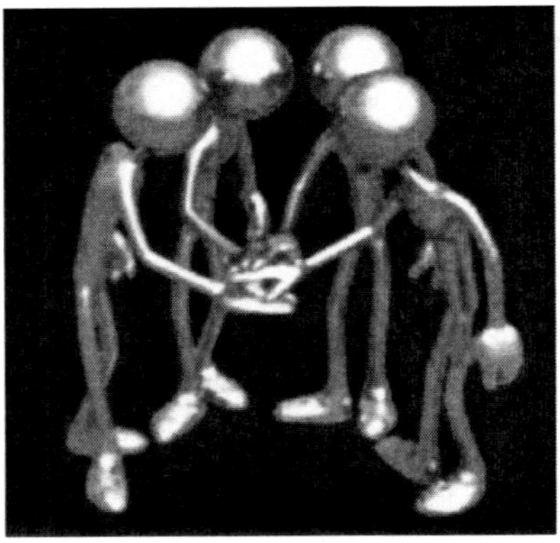

In this section you will gain insights into

1. Identifying gaps or areas of noncompliance
2. Prioritizing gap for closure
3. Stewardship of gap closure strategies

Discussion Topics

- Identifying gaps or areas of noncompliance
- Prioritizing gaps for closure
- Stewardship of gap closure strategies

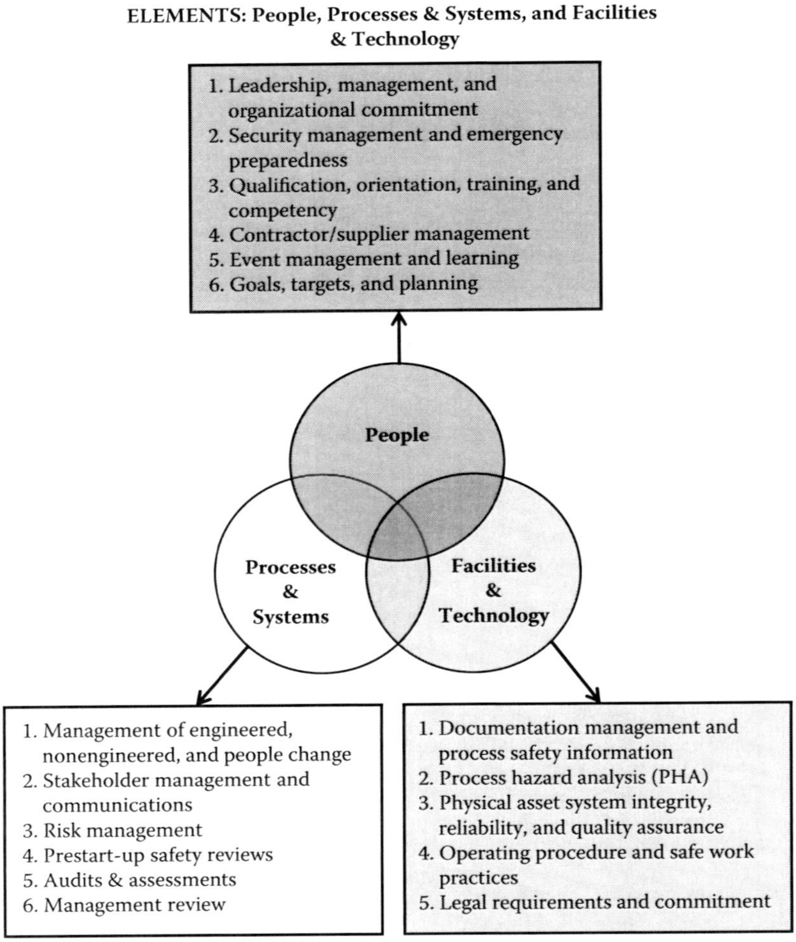
ELEMENTS: People, Processes & Systems, and Facilities & Technology
1. Leadership, management, and organizational commitment
2. Security management and emergency preparedness
3. Qualification, orientation, training, and competency
4. Contractor/supplier management
5. Event management and learning
6. Goals, targets, and planning
People
Processes & Systems
Facilities & Technology
1. Management of engineered, nonengineered, and people change
2. Stakeholder management and communications
3. Risk management
4. Prestart-up safety reviews
5. Audits & assessments
6. Management review
1. Documentation management and process safety information
2. Process hazard analysis (PHA)
3. Physical asset system integrity, reliability, and quality assurance
4. Operating procedure and safe work practices
5. Legal requirements and commitment

Organizational Management System Maturity Model

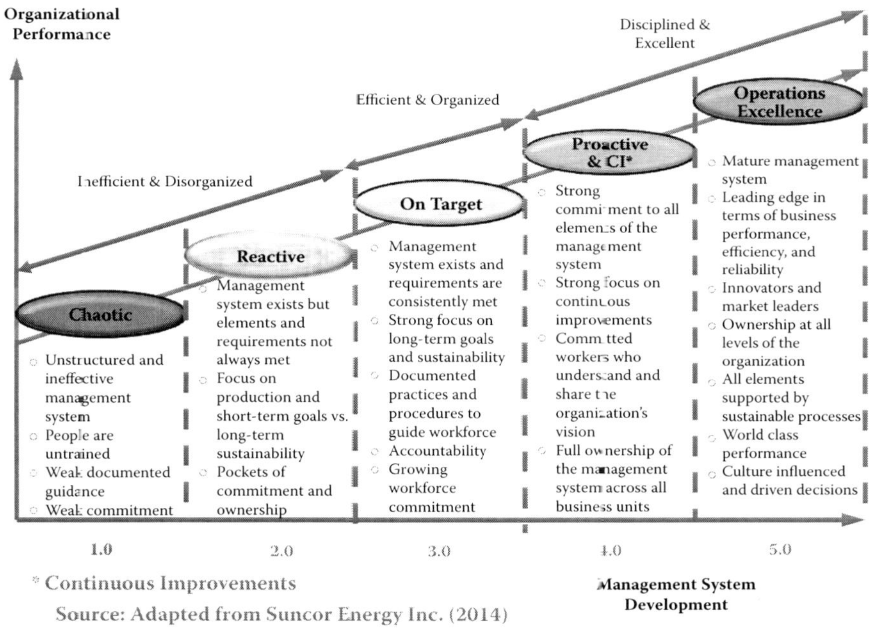

* Continuous Improvements

Source: Adapted from Suncor Energy Inc. (2014)

- ○ Each element requirement must be accompanied by an implementation guideline to ensure consistent interpretation and field application
- ○ Applicable scores are defined by the level of compliance to the element requirements
- ○ At the very minimum, the organization must strive for ***Planned*** *Compliance and Development Levels*

Identifying the Gaps

- Consider each subelement against the requirements of the implementation guidelines for compliance
- Assess where the business unit/area is relative to maturity model
- Allocate score accordingly
- Identify gap preventing full compliance ... i.e., to achieve a score of 3.0

Element	*Subelement*	*Gaps Identified*	*Avg. Score*
Management of engineered, non-engineered, and people change	o A formal MOC process must exist for managing change	o Standardized tools for managing change not available	2.0
	o Personnel must be assigned responsibilities for managing change	o Individuals identified vs. roles for assigned responsibilities	2.0
	o Authority must be in place for approving the change	o Inconsistent signoff of all required authorities for approving change	2.5
	o Infrastructure and processes must be available for communicating the change	o Very effective use of email, alerts, communication, and training tools for communicating change	3.5
	o Process must be available for evaluating the effectiveness of the change	o Inconsistent approach to managing the effectiveness of change. Documentation absent	1.5
Element Score			**2.3**

Prioritizing the Gaps

- Risk ranking the gaps identified – High/Medium/Low
  - High risk gap – addressed 1st
    - High risks of personnel injuries and fatalities, environmental degradation, and damage to facilities and reputation
- Identifying low-hanging fruits
  - Subelements requiring small incremental efforts to achieve compliance
  - A work practice may exist but process not formalized
  - Documentation as a means of achieving compliance

Closing the Gaps

- Leadership commitment
  - Removes barriers
  - Provides support
  - Resources allocation
  - Stewardship – reporting to senior leadership on an ongoing basis
- Team chartered
  - Scope of work
  - Measured targets
  - Stakeholder impact assessments
  - Project schedule defined
- Execution
- Engagement and communication

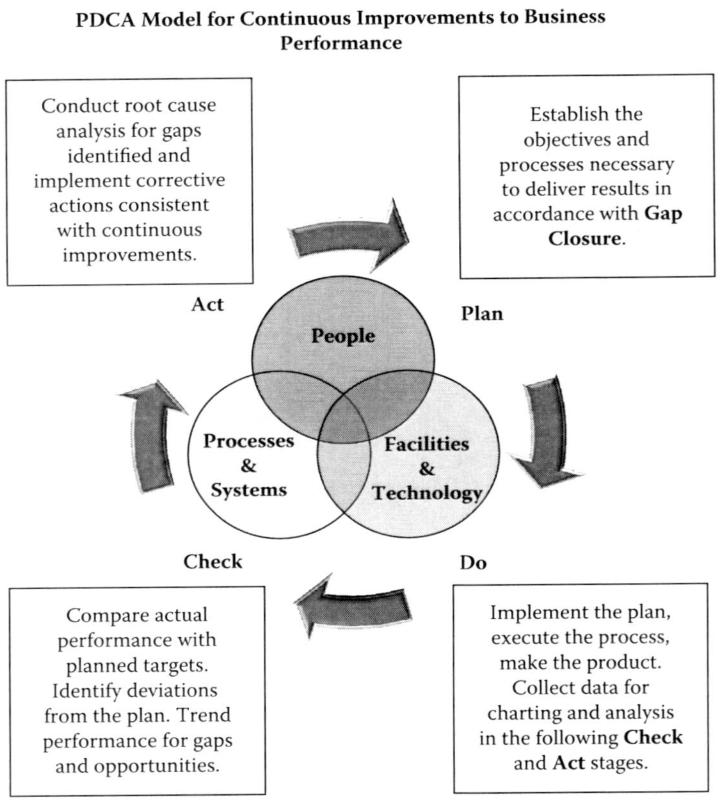
PDCA Model for Continuous Improvements to Business Performance
Conduct root cause analysis for gaps identified and implement corrective actions consistent with continuous improvements.
Establish the objectives and processes necessary to deliver results in accordance with **Gap Closure**.
Act
Plan
People
Processes & Systems
Facilities & Technology
Check
Do
Compare actual performance with planned targets. Identify deviations from the plan. Trend performance for gaps and opportunities.
Implement the plan, execute the process, make the product. Collect data for charting and analysis in the following **Check** and **Act** stages.

Fundamental 7

Continuous Improvements and Shared Learning – Networks and Communities of Practice

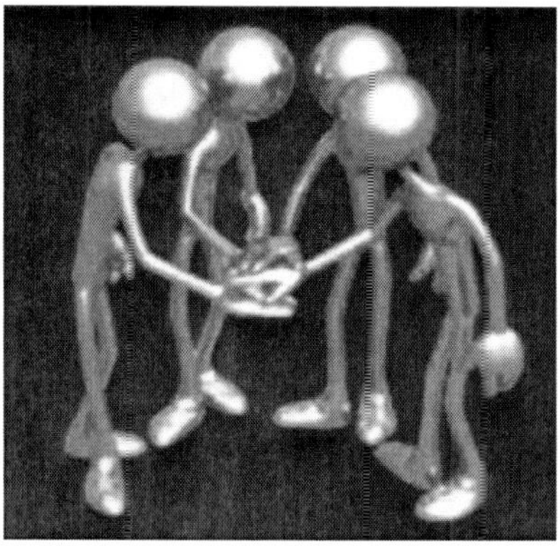

In this section you will gain insights into

1. Networks development
2. Incident reporting
3. Incident investigation
4. Organized sharing of learning

Discussion Topics

- Creating and activating networks
  - Subject matter experts (SMEs)
  - Focused scope of work
  - Technical support
  - Community of practice
- Incident reporting system
- Incident investigation model
- Shared learning
  - Leadership challenge
  - Engineering to operations

Elements: People, Processes, & Systems, and Facilities & Technology

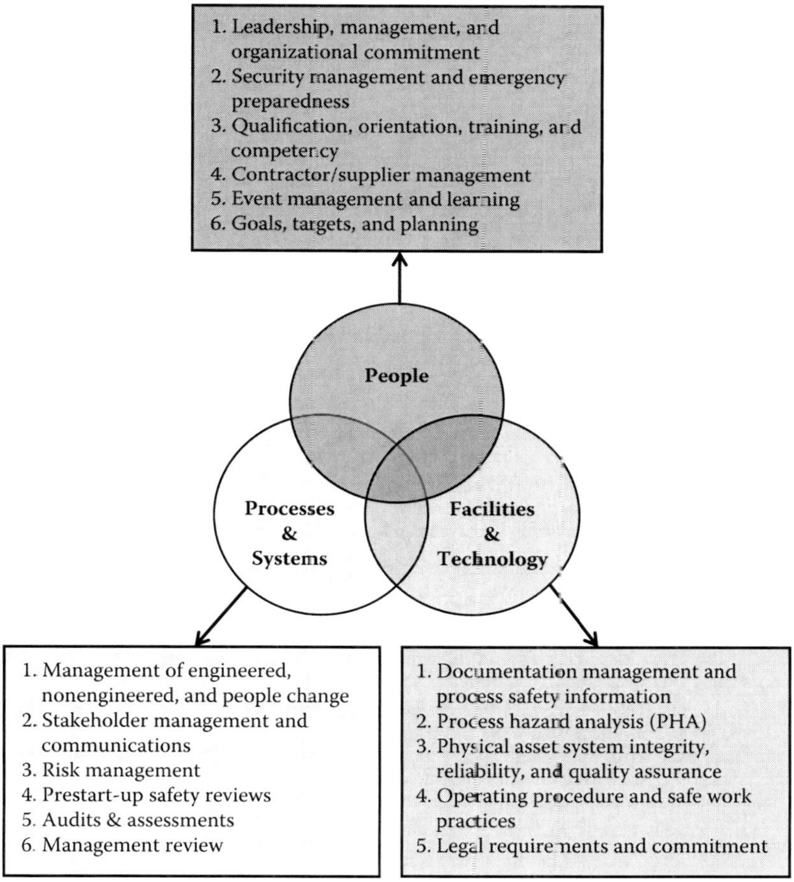

Creating and Activating Networks

Objectives
Continuous improvements in performance management of elem ent

Network Requirements

- o Clearly defined scope of work
- o 3–5 core team members
- o Business areas/stakeholder representation
- o Network charter
- o Support services – communications, legal, etc.
- o Leadership support

Network Composition

- o Strong, credible leader
- o Subject matter experts
- o Committed, motivated team members
- o Extended resource pool to be drawn from as required

Network Operation

- o Key deliverables defined and prioritized
- o Avenue created for knowledge and information transfers from the frontline to the network
- o Knowledge creation process – best practice identification and transfer to the frontline

Network Structure

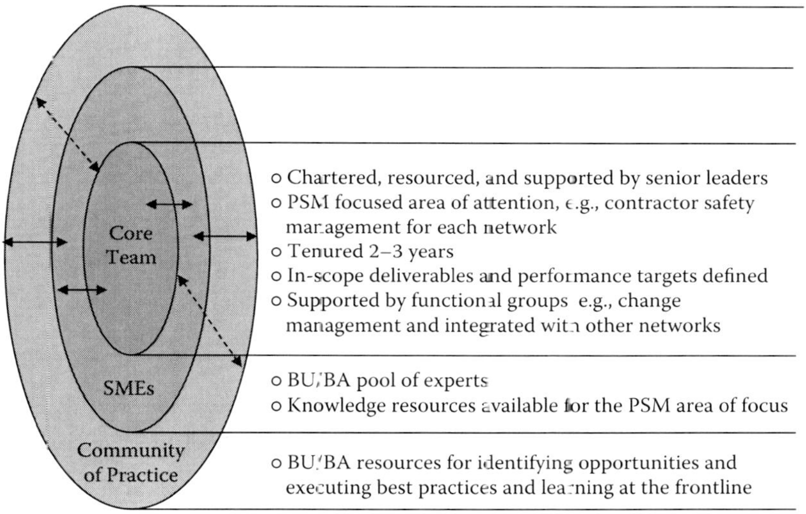

Source: Safety Erudite Inc. (2012)

- Each network should comprise 3–5 core team members
    - A strong respected leader with technical and leadership skills required
- Subject matter experts (SMEs) should be identified from each BU/BA to ensure ownership
- The community of practice forms the conduit for bringing new opportunities to the organization and transferring new processes to action at the frontline

Challenges to Getting Knowledge to the Frontline

- Leadership capabilities
- Fear of legal/market responses
- Weak understanding and communication of the benefits of shared learning
- Absence of the machinery within the organization for generating learnings
- Absence of an organized method for sharing – cost issues

Essentials for Effective Shared Learnings

- Learnings should be simple and easy to understand and apply
- Learnings should be repeatable
- Medium and technology for sharing learnings should cater to both Generations X & Y
- Where learning from incidents is concerned, focus should be on the following:
    - What happened?
    - Root causes of incidents
    - Key learnings from incidents
    - Recommendations to prevent incidents from being repeated
- A corporate approach to capturing and sharing learning
- An expert networks for generating continuous improvements (Experts/SMEs)
- Focus on proactive measures
- Have a model for generating knowledge and learning
- Develop a model for sharing knowledge and learning
- Tools for transferring knowledge and learnings to the frontline
- Use standardized templates and processes for sharing – alerts/investigation summaries/best practices
- Establish an organizational process approval and for controlling quality
- Action management stewardship and follow-up
- Sharing tools must be user friendly, searchable, accessible to all workers, accommodating to collaboration, and secure

Model for Generating Knowledge and Learning

- Incident reporting and management system
- Assess risk and consequences
- Incident investigation and Root Cause Analysis (RCA)
- Action management and follow-up

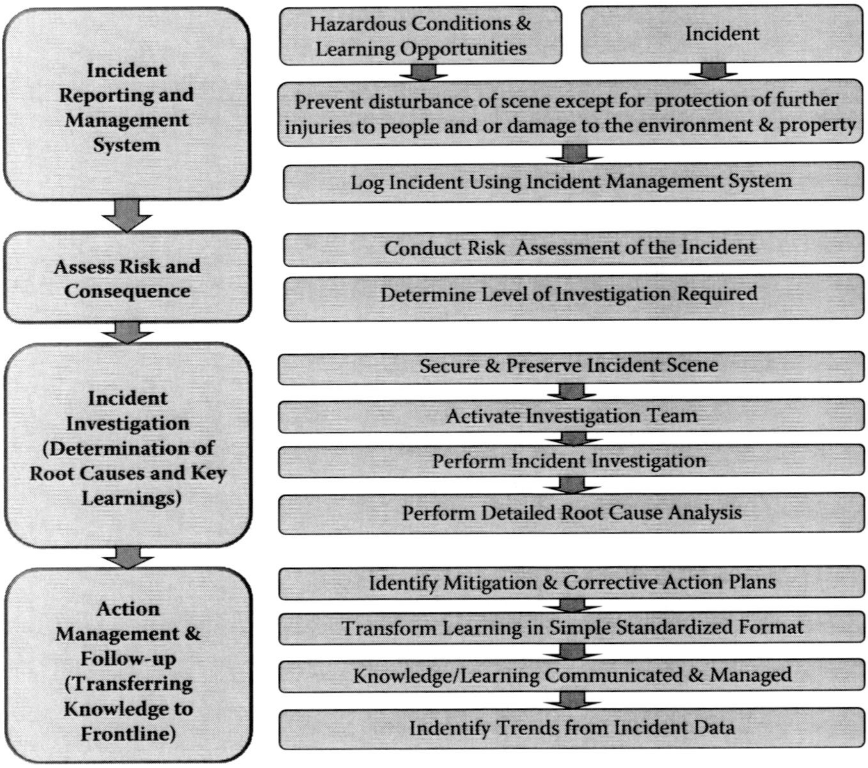

Value–Effort Relationships – Learnings

- Executing learnings requires identification of the value–effort relationships
- Nice to have – High Effort/High Value
- Best practices – Low Effort/High Value

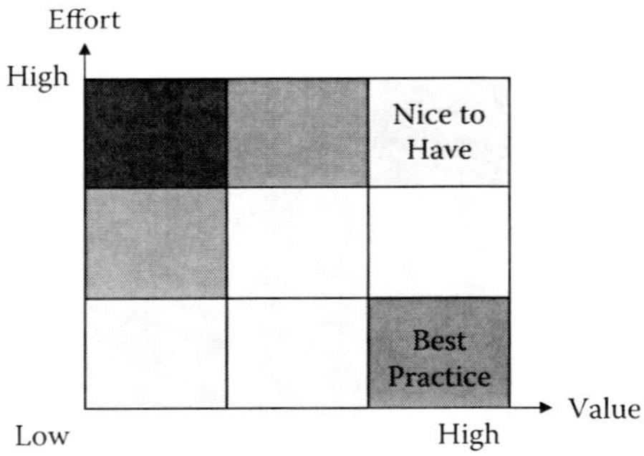

Opportunity Matrix for Prioritizing Work

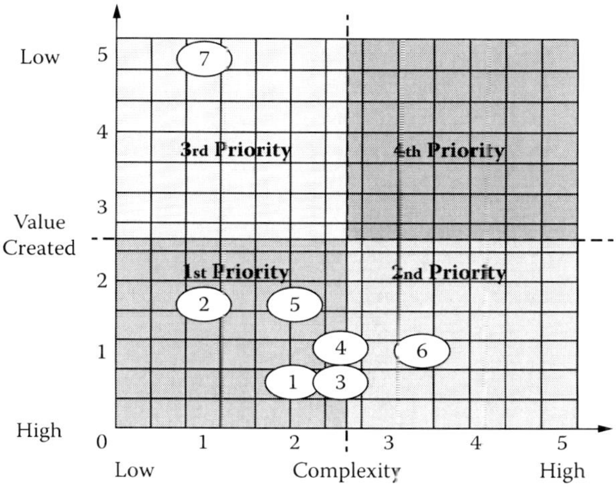

Priority	Opportunitiy	Complexity 1=Low 5=High	Value Created 1=High 5=Low
1	Get prequalification rolled out right	2.0	0.5
2	Align contract language with PSM/OMS requirements	1.0	1.5
3	Standardized computer-based contractor orientation (3 levels)	2.5	0.5
4	Simplify CM101 and make BU/BA specific (workshop with tools)	2.5	1.0
5	CM 101 supposed to BU in rollout of Contractor Safety	2.0	2.0
6	Provide access to organization's work-related information (procedures, policies, processes) to all contractors	3.5	1.0
7	Update the contractor safety standard based on improvement oppportunities identified	1.0	5.0

Model for Sharing Learnings

- Executing learnings requires identification of the value–effort relationships
- Nice to have – High Effort/High Value
- Best practices – Low Effort/High Value

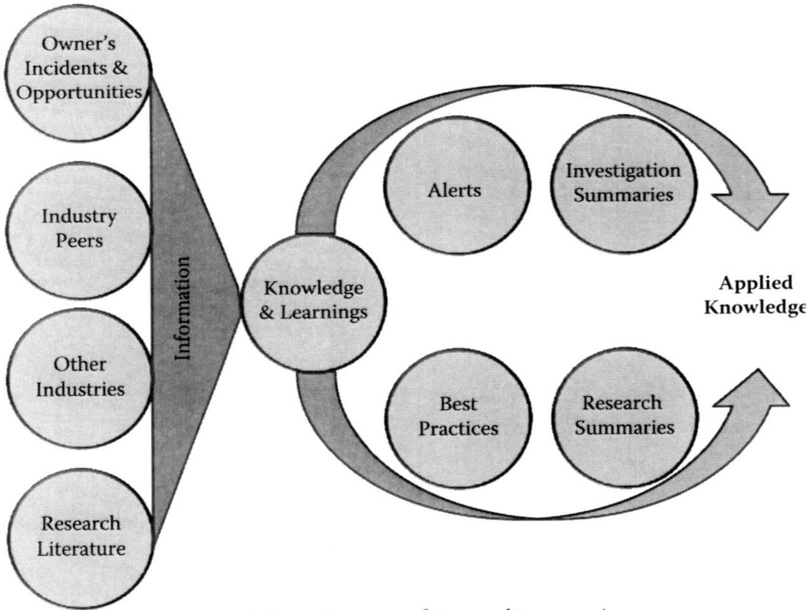

https://www.safetyerudite.com/

Shared Learning – Current Situation

Sharing of Knowledge	Market Situation
Within Organization – Internal (Business Units and Functional Areas)	***Early Stages of Development*** o Organizations are in the early stages of assessing the value of shared learning o Now evaluating how to share knowledge more effectively o No standardized processes for sharing of knowledge o Emphasis continues to be on sharing of data and information as opposed to knowledge that is beneficial to organizations
Within Industry	***Disorganized & Indiscriminate*** o Sharing within industries is disorganized and indiscriminate o Information is shared informally among peers via email in a disorganized manner with no concerns of the type and quality of information being shared o Very little regard for sensitivity of the information being shared as well as the accompanying liability in sharing
Across Industries	***Not Available*** o Very little, if any, sharing of knowledge occurs across industries

Shared Learning – Leadership Essentials

- Leadership must provide learning in a format that is structured and presented to cater to both Generations X & Y
  - Both groups learn differently and their learning needs must be met for success
- Leadership must seek to proactively pull learning from various sources such that learning can make it to the frontline where incidents occur
  - Internal sources
  - From peers
  - Sources such as research and investigations studies
- Leadership must recognize the value of doing work right once to reduce cost and improve overall performance
  - Shared learning has the potential to significantly reduce the number of workplace incidents and associated costs
- Bold and transparent leadership behaviors that challenge the status quo are required
- Leading and managing change is a core competency of leaders in order to be successful in transforming organizations to learning cultures

Required Transformational Leadership Behaviors

- Creating a shared vision
- Promoting involvement, consultation, and participation
- Creating an organizational environment that encourages creativity, innovation, proactivity, responsibility, and excellence
- Having moral authority derived from trustworthiness, competence, sense of fairness, sincerity of purpose, and personality
- Leading through periods of challenges, ambiguity, and intense competition or high-growth periods
- Promoting intellectual stimulation
- Considering the individual capabilities of employees
- Willing to take risks and challenge the status quo
- Leading across cultures and international borders
- Building strong teams while focusing on macro-management
- Being charismatic and motivating workers to strong performance

Shared Learning – Leadership Challenges

- Lack of desire to challenge the status quo
- Apathy: It's not broken, no need to fix it
- Leadership skills deficiencies in leading and managing change
- In all instances, the following considerations are essential in change management:
  - A clear definition of the desired change
  - Stakeholders identified and mapped
  - Stakeholder impact assessment
  - Engagement and involvement of stakeholders
  - Change impact by category
    - Roles and responsibilities
    - Process
    - Technology
    - Skills and knowledge
  - Timing of change – When does change occur
  - Support in executing change
    - Communication
    - Training
    - Leadership alignment
  - Risk associated with executing the change
- Leadership focus on silo performance and profit maximization as opposed to value maximization enterprise-wide
- Absence of a shared corporate vision that has permeated the entire organization

Remember the ABCD Model for Building Trust

ABLE – Demonstrate Competence

- Produce results
- Make things happen
- Know the organization/set people up for success

BELIEVABLE – Act with Integrity ... Be Credible

- Be honest in dealing with people; be fair/equitable/consistent /respectful
- Value-driven behaviors reassure employees that they can rely on their leader

CONNECTED – Demonstrate Genuine Care and Empathy for People

- Understand and act on worker needs/listen/share information /be a real person
- When leaders share a little bit about themselves, it makes them approachable

DEPENDABLE – Follow Through on Commitment

- Say what you will do and do what you say you will
- Be responsive to the needs of others
- Be organized and reassure followers

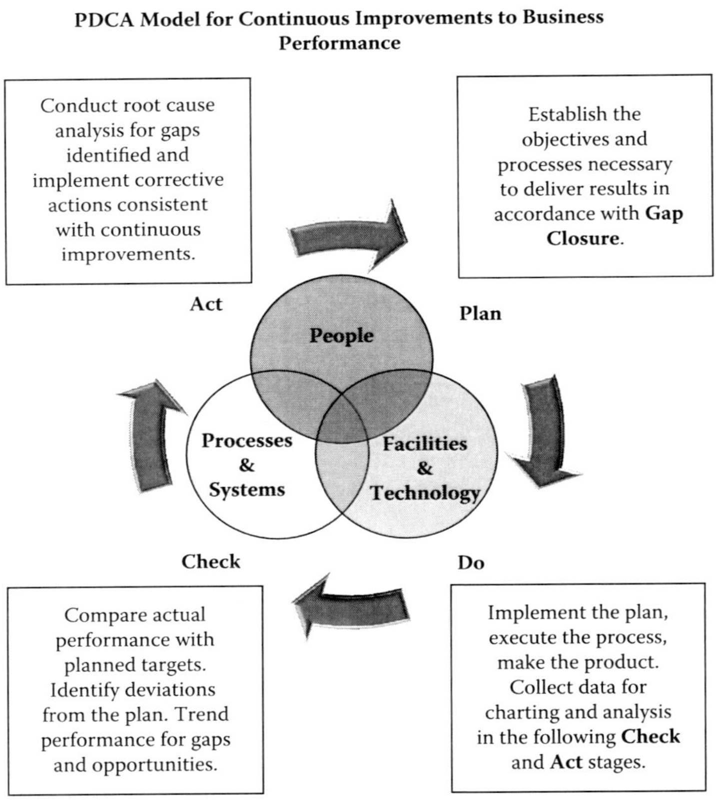
PDCA Model for Continuous Improvements to Business Performance
Conduct root cause analysis for gaps identified and implement corrective actions consistent with continuous improvements.
Establish the objectives and processes necessary to deliver results in accordance with Gap Closure.
Act
Plan
People
Processes & Systems
Facilities & Technology
Check
Do
Compare actual performance with planned targets. Identify deviations from the plan. Trend performance for gaps and opportunities.
Implement the plan, execute the process, make the product. Collect data for charting and analysis in the following Check and Act stages.

Summary

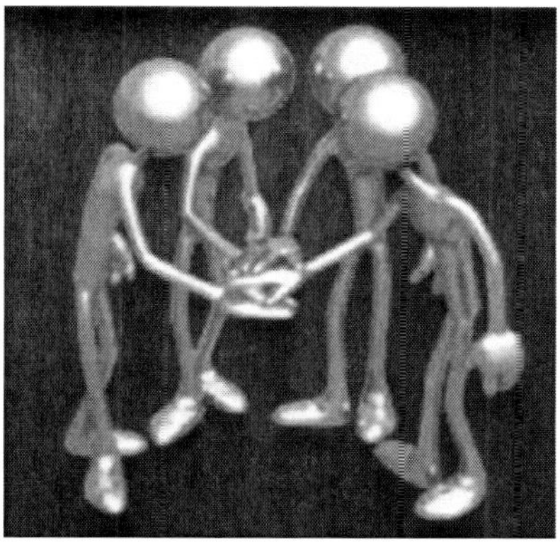

OEMS – A Definition

An Operationally Excellent Management System is one that

- Develops its people to high levels of discipline and competence
- Is supported by adequate tools and processes, adjusts to the continually changing business environment, and is guided by policies, standards, and procedures for ensuring the integrity of its assets
- Provides standardization for global operations
- Prioritizes the management of the environment, health, and safety to continually improve the reliability and efficiency of business performance
- With strong leadership commitment, caters for adopting best practices and standards that help in the delivery of world-class performance and sustained value maximization
- Such systems demonstrate sustained exceptional performance and are noted for few incidents and preservation of the environment

Operational excellence is not something separate from our business; it is how we must run our business to achieve our vision of success.

– Chevron

Benefits of an OEMS

Peer and Industry Recognition

- Knowhow and competence
- Best practices
- Goodwill

Organizational Performance

- Standardized processes
- Operational discipline
- Strong reliability
- Organized and planned performance
- Output and financial performance
- Excellent environmental performance
- Performance development
- Motivated workforce
- Little turnover

The purpose of business is to create and keep a customer.
– Peter Drucker

Elements: People, Processes, & Systems, and Facilities & Technology

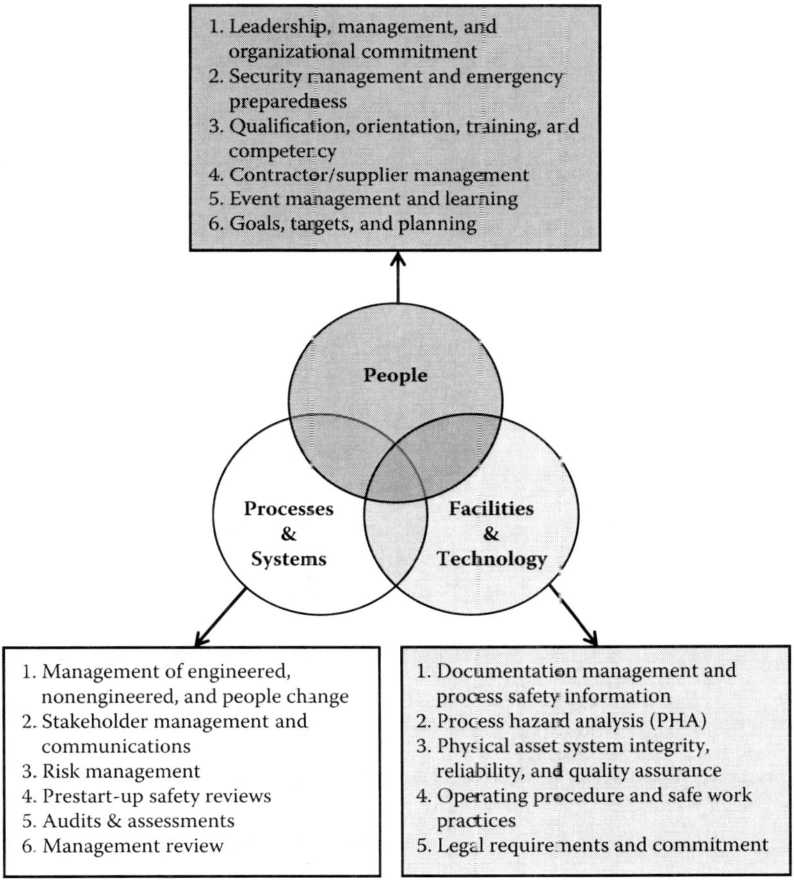

Leadership Expectations of the Future

- Increasing numbers of Gen Y in the workplace ...
  - ➢ Master of my destiny – INVOLVE ME
  - ➢ Little field experience – PROTECT ME
  - ➢ Flexible hours and locations – REMOTE ME
  - ➢ I want to grow quickly and take charge – PROMOTE ME
  - ➢ I want to take charge – EMPOWER ME

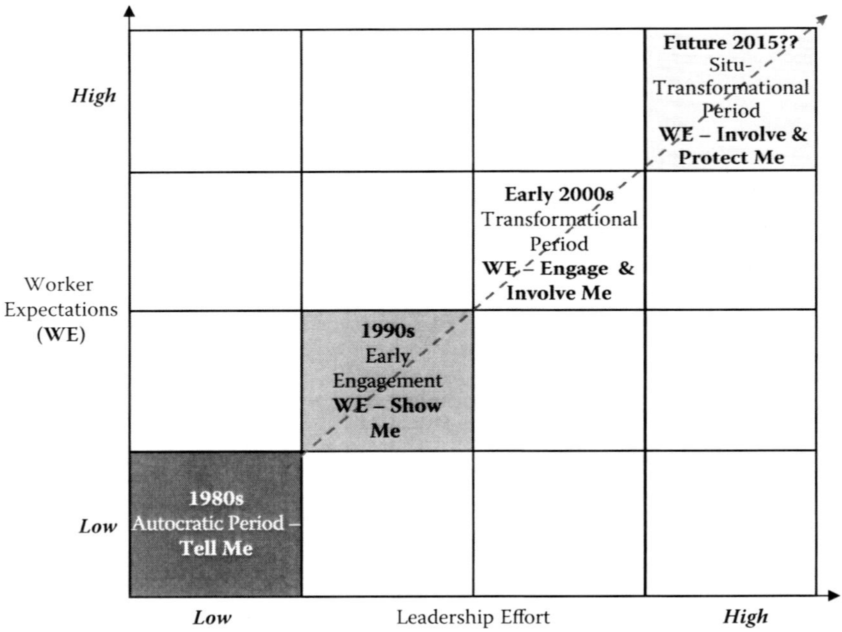

What We Say and What Others Hear

True Message	Organizational Level	Conflicting Messages
Safety Is Our Top Priority	**CEO & Senior Leadership**	o Market and shareholders reward profits o CEO loses job when profits are lower than peers but safety record is better o Not willing to recognize a lag in profits between investments in safety and long-term sustainable profitability
Safety before Production	**Middle Managers**	o Production rewarded over safety o Down time and outage frowned upon – correcting safety deficiencies requires down time and outage o Preventive maintenance often deferred based on pricing and profit drivers o Demotions and job loss as an outcome of lower than peer production performance o Weak recognition of a lag between sustained productivity and investment in safety
Safety First	**Frontline Supervisors and Workers**	o Frontline supervisors poorly trained o Senior and frontline leaders talk safety but do not demonstrate the behavior...why should I? o Getting the job done quickly o Nondetected shortcuts rewarded
Safety First	**Contractors**	o Bid process rewards lowest prices – safety budgets generally the first to be chopped to be competitive o Getting the job done quickly rewarded o Contractors often provided the dirtiest, difficult, and dangerous jobs o Punitive consequences for reporting incidents and near misses drives reporting underground leading to lost learning opportunities

Source: Lutchman, Maharaj, & Ghanem (2012)

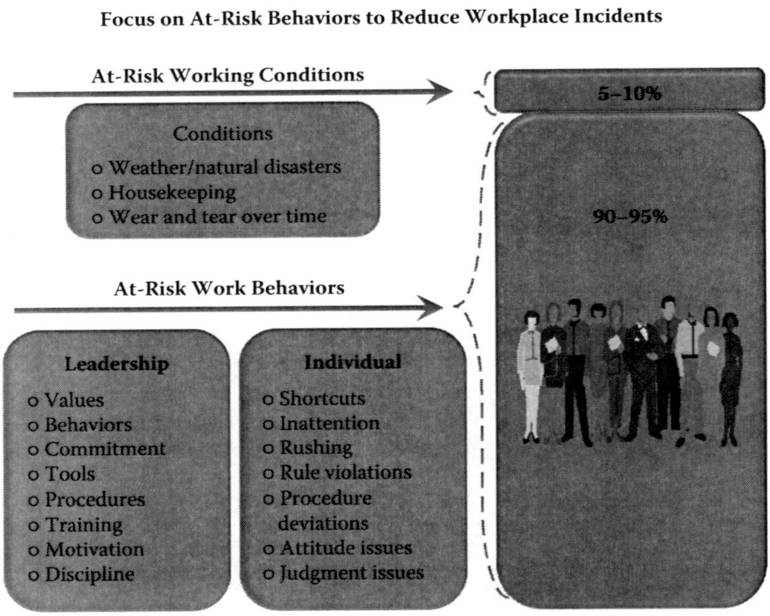
Focus on At-Risk Behaviors to Reduce Workplace Incidents
At-Risk Working Conditions
Conditions
o Weather/natural disasters
o Housekeeping
o Wear and tear over time
At-Risk Work Behaviors
Leadership
o Values
o Behaviors
o Commitment
o Tools
o Procedures
o Training
o Motivation
o Discipline
Individual
o Shortcuts
o Inattention
o Rushing
o Rule violations
o Procedure deviations
o Attitude issues
o Judgment issues
5–10%
90–95%

Management System Disaggregated

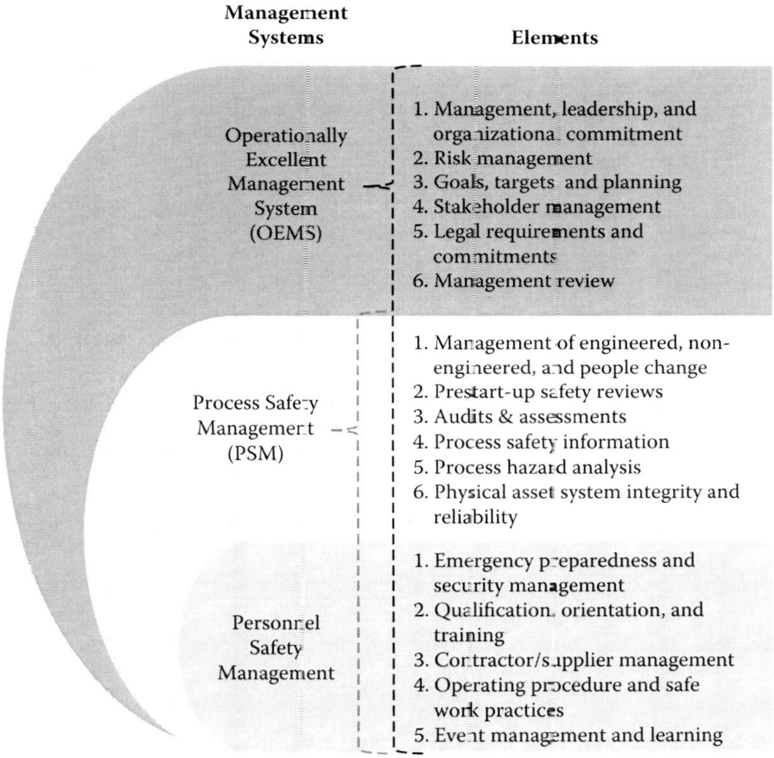

Safety Management System = Personnel Safety Management

\+

Process Safety Management

OEMS = SMS on **Steroids**

People Element

3. Qualification, Orientation, and Training & Competency

Training Method Effectiveness

- On-the-job training, mentoring, and coaching is most effective
- Involvement and engagement
- Showing in the field

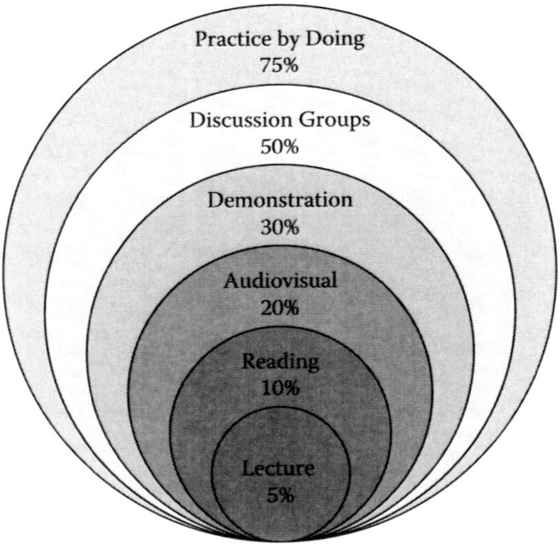

People Element

4. Contractor/Supplier Management

Contractor Safety Audits Highly Beneficial

- Contractors want to do a good job and protect their workforce
- Contractors want to work collaboratively with owners to identify and close gaps in their safety management systems
- Collaborative audits improve EH&S performance

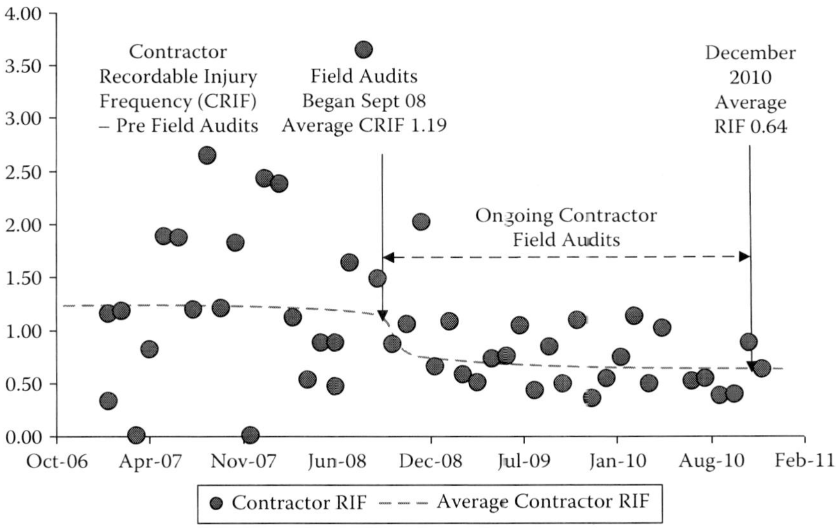

Source: Lutchman, Maharaj, & Ghanem (2012)

People Element

5. Event Management and Learning

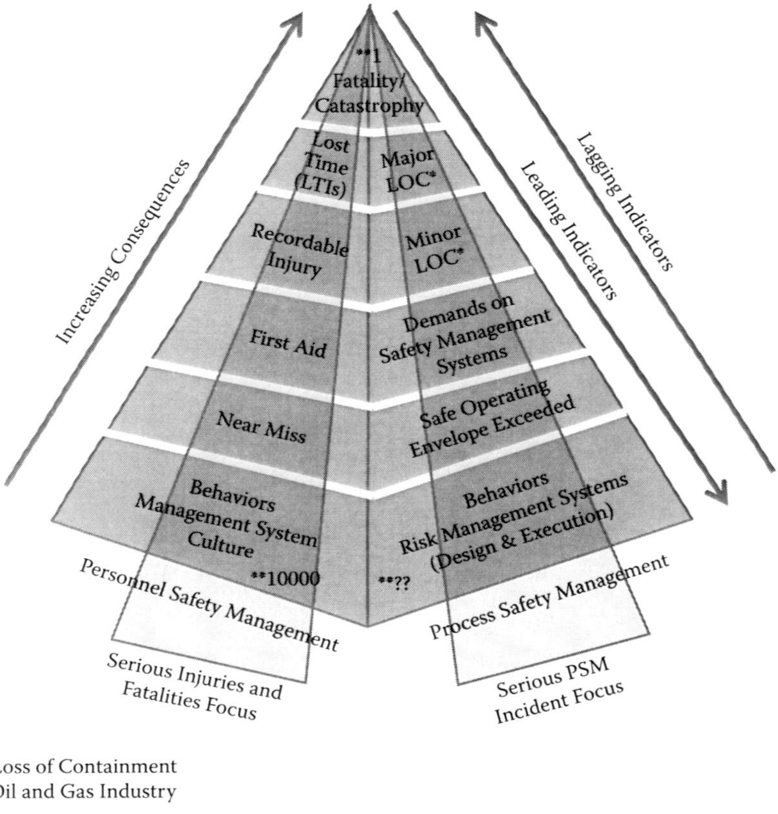

*Loss of Containment
**Oil and Gas Industry

Processes & Systems Element

1. Management of Engineered, Nonengineered, and People Change (MOC)

Layers of Protection

- Leverages the learnings of the Swiss Cheese Model
- 1st Layer – Design out the risks or hazards
- 2nd Layer – Engineered controls
- 3rd Layer – Administrative controls
- 4th Layer – LAST LINE OF DEFENSE – PPE

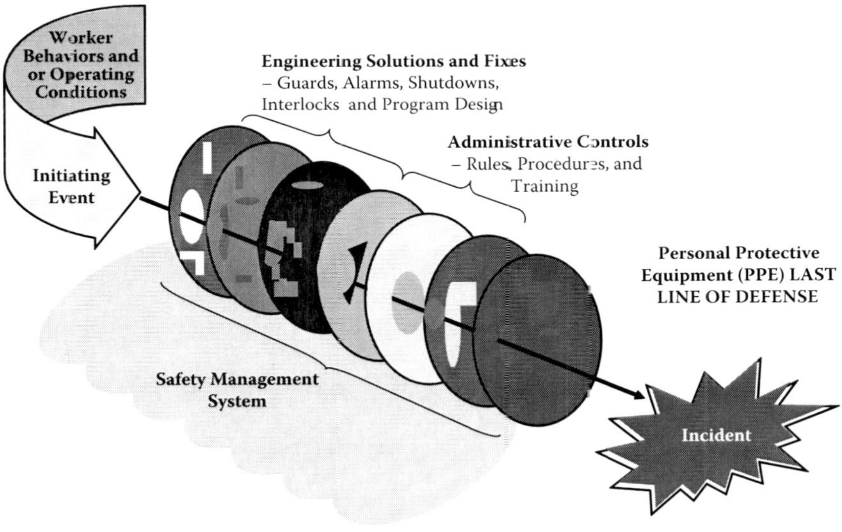

Source: Lutchman, Maharaj, & Ghanem (2012)

Processes & Systems Elements

2. **Stakeholder Management**

Stakeholder Interests Varies

- Create a stakeholder interest map
- Understand each stakeholder interest
- Engage and communicate
- **Not every stakeholder needs to be satisfied ... but they all want to feel considered and engaged**

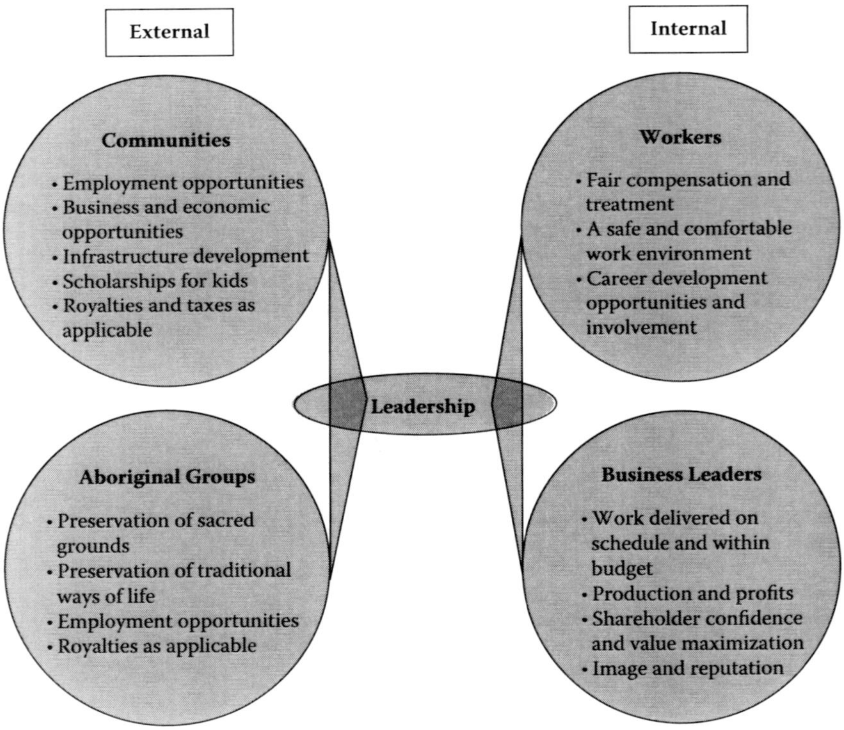

Processes & Systems Elements

6. Management Reviews

Establish and Steward SMART Performance Indicators

Leadership establishes the requirements reporting on Key Performance Indicators (KPIs) to ensure continuous improvements and to assess the efficacy of processes and tools that may be in place to support the OMS

1. Establishing KPIs
   - ➢ Leading and lagging indicator focus
   - ➢ Emphasis on leading indicators
   - ➢ Monthly reporting and stewardship

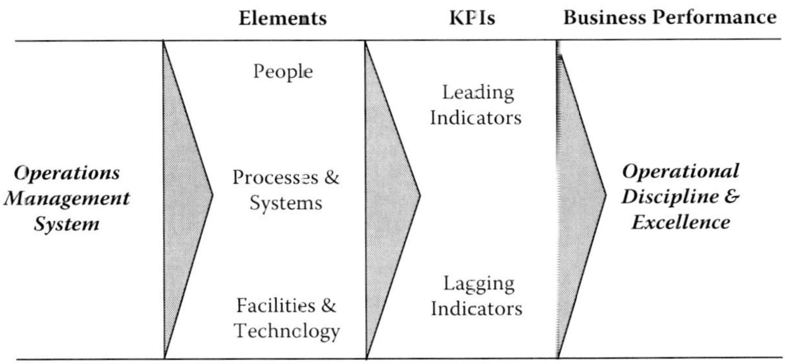

Processes & Systems Elements

6. **Management Reviews**

Operations Management System

People Elements

Leadership, management, and organizational commitment

Security management and emergency preparedness

Qualification, orientation, and training

Contractor/supplier management

Event management and learning

Goals, targets, and planning

Facilities & Technology Elements

Documentation management and process safety information

Process hazard analysis (PHA)

Physical asset system integrity and reliability

Operating procedure and safe work practices

Legal requirements and commitment

Processes & Systems Elements

Management of engineered, non-engineered, and people change

Stakeholder management and communications

Risk management

Prestart-up safety reviews (PSSR)

Audits & assessments

Management review

Approaches to KPIs

- o Identify 3–5 KPIs for each element for stewardship
- o Select high-value KPIs to focus on
- o Focus on a maximum of 2–3 KPIs to steward
- o Ensure SMART capabilities for each KPI
- o Determine the right frequency for reporting on KPIs – monthly vs. quarterly
- o Avoid taxing the business in reporting requirements

Capabilities of the Frontline Supervisor

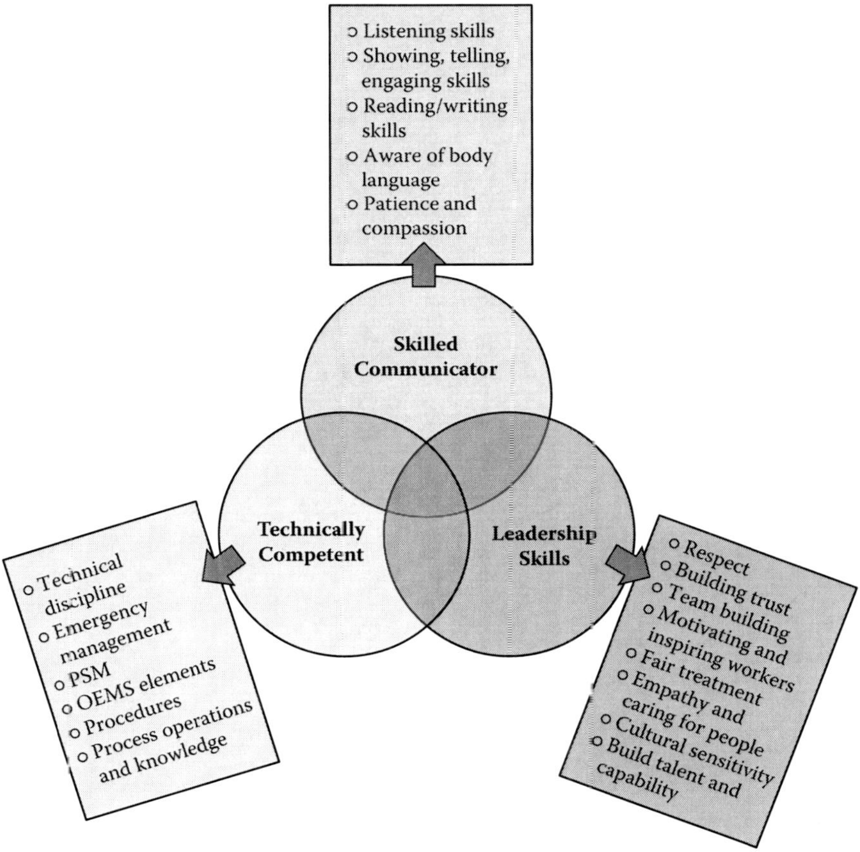

Situ-Transformational Leadership Model

- Combines transformational leadership behaviors and Situational Leadership Model
- Transformational leadership behaviors applied at each stage of the worker's development

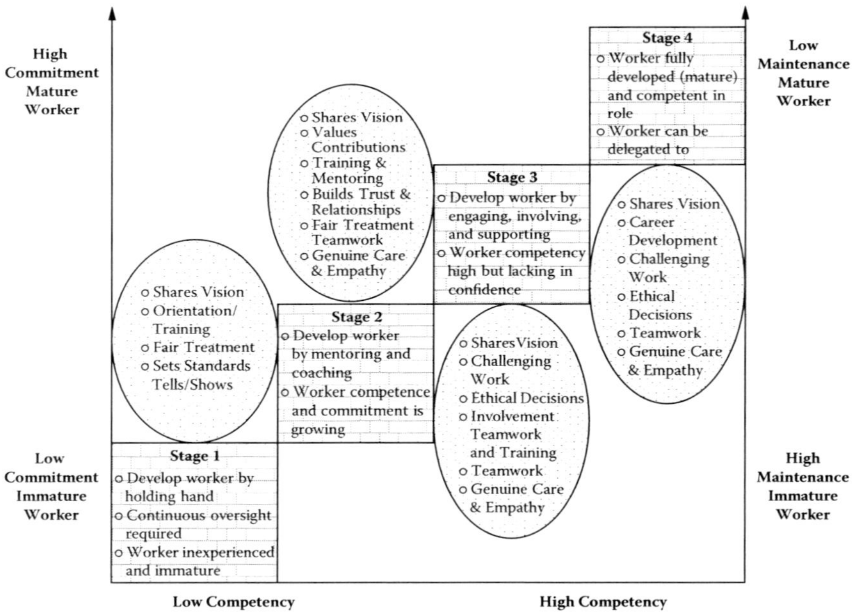

Source: Lutchman, Maharaj, & Ghanem (2012)

Key Leadership Focus for OEMS

- Sharing the vision
    - Senior leadership communication
    - Town hall meetings
    - Engagement sessions
- Managing change
    - Identifying and engaging stakeholders
    - Stakeholder impact assessment
        - Business unit
        - Business area
        - Departmental
        - Individual
- Genuine care & empathy
    - Changing work loads
    - Different ways of doing things
    - Personnel impact
- Teamwork
    - Senior leadership communication
    - Town hall meetings
    - Engagement sessions
- Developing workers
    - Directing
    - Supporting
    - Coaching
    - Delegating

Understanding the Element Compliance Process

Element
Event Management and Learning

Subelement Requirements

- o Incidents and hazards reporting
- o Incidents and hazards risk assessments
- o Incident investigations and root cause analysis
- o Corrective actions management
- o Knowledge generated and Learning from Events (LFE) process

Implementation Guidelines

- o Consistent interpretation of each subelement requirement

Subelement Interpretation

- o Incidents and hazards reporting – A process/tool is in place for all personnel to report and record hazards and incidents. Personnel are trained in use of the process/ tool. The process/tools provides a notification for all high-risk incidents and hazards with appropriate tracking mechanisms to ensure action management and follow-up

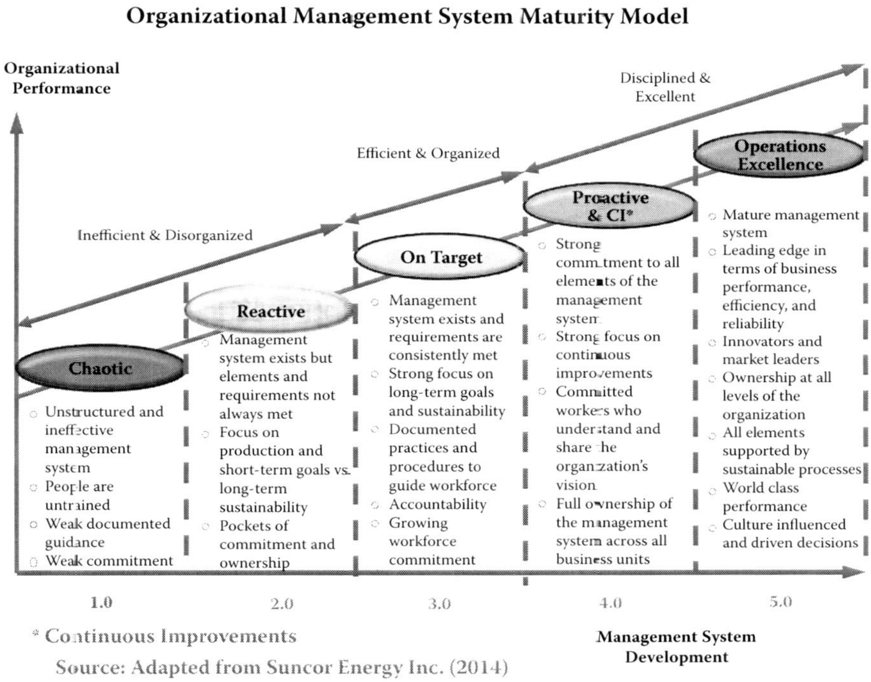

Source: Adapted from Suncor Energy Inc. (2014)

- Each element requirement must be accompanied by an implementation guideline to ensure consistent interpretation and field application
- Applicable scores are defined by the level of compliance to the element requirements
- At the very minimum, the organization must strive for ***Planned*** *Compliance and Development Levels*

Steps in Establishing the Baseline

Planning for Establishing the Baseline ~ 2 Weeks

- o Communication – Awareness
- o Leadership commitment – Selling the value/resources requirements/ developing the scope of work
- o Training – Use of supporting tools for information gathering
- o Done for each BU

Documentation Mapping and Information Gathering ~ 2 Weeks

- o Data gathering to support compliance to element requirements
- o Check-sheets/procedures
- o Documented processes
- o Clear and consistent understanding of element requirements and scoring
- o Subject matter expert (SME) – element lead identified

Verification ~ 1 Week

- o Cold-eyes review and verification of supporting documentation and scoring
- o Is it reflected in the field?
- o Enlist expertise of auditors
- o SME/auditor verifies scoring

Implementation Plan Development ~ 2 Weeks

- o Gaps identification
- o Identification of high-risk exposures and low-hanging fruits
- o Prioritization
- o Gap closure strategy – Resourcing and implementation
- o Operational discipline

Sustainment – Plan–Do–Check–Act (PDCA Model)

PDCA Model for Continuous Improvements to Business Performance

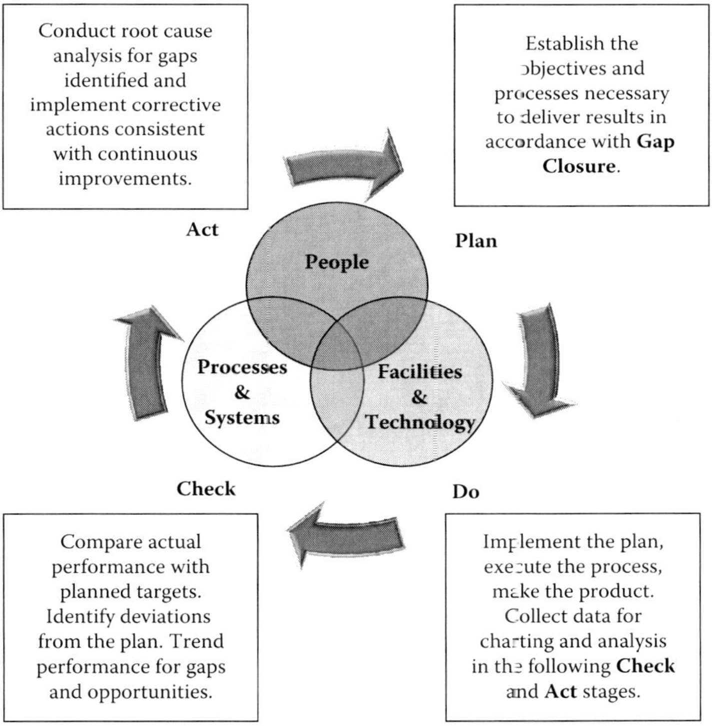

Auditing from the Baseline or Prior Audit

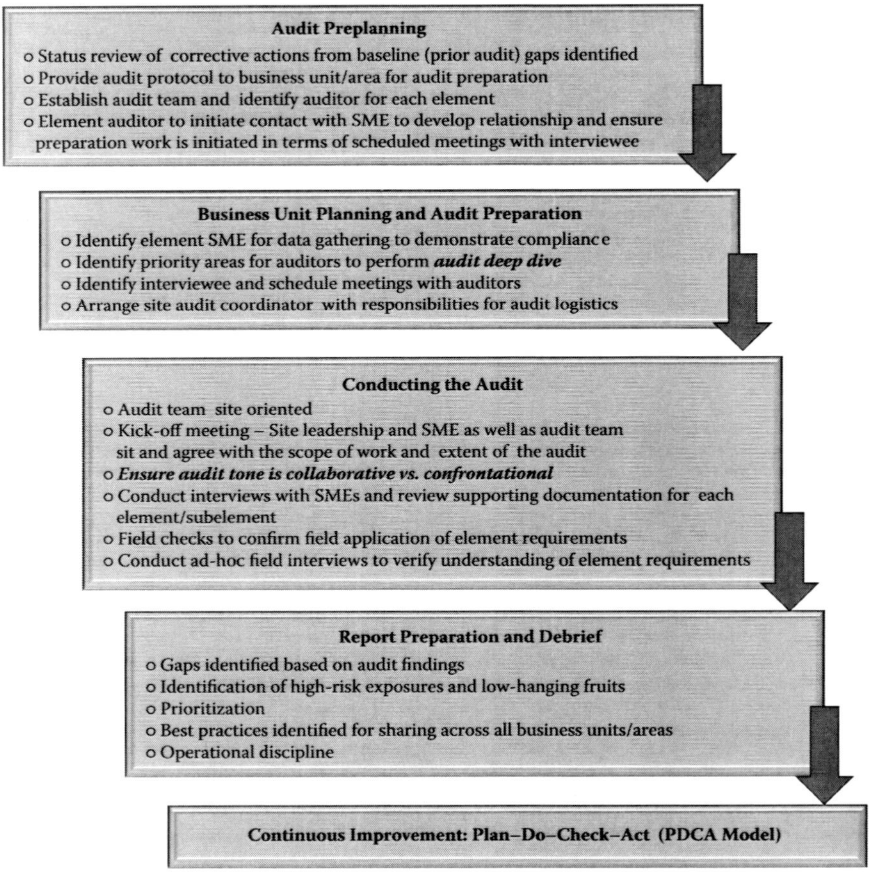

Prioritizing the Gaps

- Risk ranking the gaps identified – High/Medium/Low
    - High-risk gap – addressed 1st
        - High risks of personnel injuries and fatalities, environmental degradation, and damage to facilities and reputation
- Identifying low-hanging fruits
    - Subelements requiring small incremental efforts to achieve compliance
    - A work practice may exist but process not formalized
    - Documentation as a means of achieving compliance

Closing the Gaps

- Leadership commitment
    - Removes barriers
    - Provide support
    - Resources allocation
    - Stewardship – Reporting to senior leadership on an ongoing basis
- Team chartered
    - Scope of work
    - Measured targets
    - Stakeholder impact assessments
    - Project schedule define
- Execution
- Engagement and communication

Network Structure

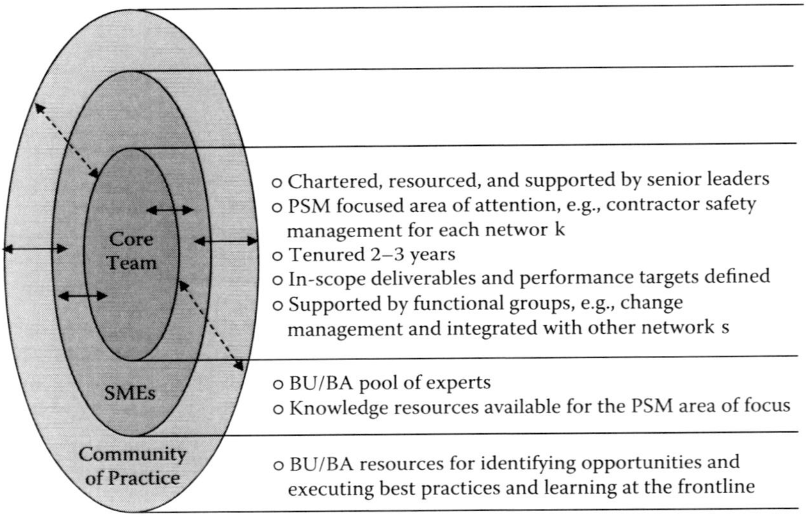

Source: Safety Erudite Inc. (2012)

- Each network should comprise 3-5 core team members
  - A strong respected leader with technical and leadership skills required
- Subject matter experts (SMEs) should be identified from each BU/BA to ensure ownership
- The community of practice forms the conduit for bringing new opportunities to the organization and transferring new processes to action at the frontline

Model for Generating Knowledge and Learning

- Incident reporting and management system
- Assess risk and consequences
- Incident investigation and Root Cause Analysis (RCA)
- Action management and follow-up

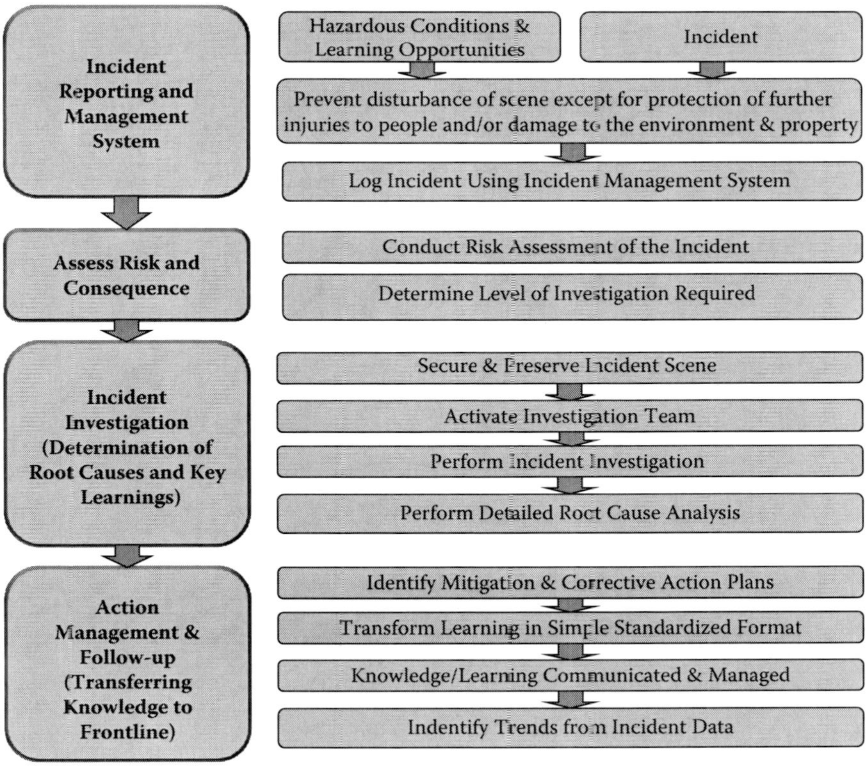

Model for Sharing Learnings

- Executing learnings requires identification of the value–effort relationships
- Nice to have – High Effort/High Value
- Best Practices – Low Effort/High Value

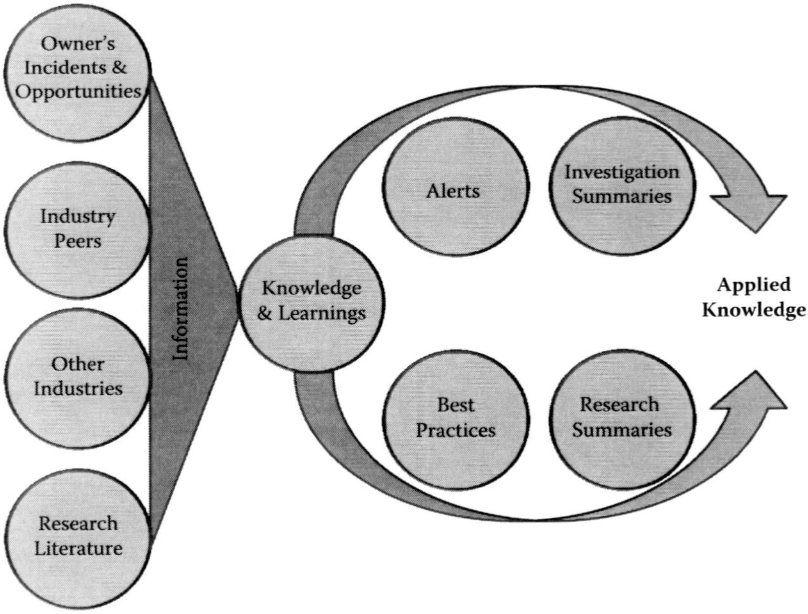

Index

I

J

K

L